History and Evolution of Technology
COUNTING

(1st in the series)

Jesse R Luckett

This book series has been developed with the assistance of artificial intelligence (AI) tools for historical research, content generation, fact-checking, and editorial validation. While great care has been taken to ensure the accuracy, clarity, and scholarly integrity of the material presented, AI-assisted research is not infallible and may reflect the limitations of its data sources or interpretations.

Readers should note that historical interpretation often involves nuance, context, and evolving scholarly consensus. As such, this work should not be considered a substitute for direct engagement with primary sources or peer-reviewed academic studies. Where possible, citations and references have been included to encourage further exploration and independent verification.

The authors and publishers disclaim any liability for inaccuracies, omissions, or consequences arising from the use or reliance on the content herein. Readers are encouraged to apply critical thinking and consult additional resources when engaging with historical and technological subjects presented in this series.

Printed in the United States of America

First Edition, August, 2025

ISBN: 978-1-968685-06-5

RL
www.jesseluckett.com

To my amazing and supportive family, including the incredibly gifted grandkids, Oliver and Bijan! Such a proud parent and grandparent I am!

And to everyone, never stop learning.

Table of Contents

Preface

This book embarks on an exciting journey, tracing the captivating evolution of counting through the dawn of computation from its humble origins in the minds of early humans to the sophisticated digital systems that define our modern world. We will explore the ingenious methods employed by our ancestors to quantify their surroundings, from the simple act of using fingers to represent quantities to the development of complex written numerical systems. The narrative will explore the fascinating stories of ancient civilizations and their unique approaches to counting, showcasing the creativity and ingenuity of human thought in tackling this fundamental problem. We will examine the development of various numeral systems, comparing and contrasting their strengths and limitations, and revealing how they shaped the course of mathematics and science. Along the way, we will encounter pivotal moments in the history of computation, such as the eventual invention of zero, the creation of mechanical calculators, and the advent of electronic and digital computers. Each innovation will be contextualized, illustrating not only its technical significance but also its wider social and cultural implications. The narrative will be accessible to a wide audience, avoiding complex mathematical jargon whenever possible, yet offering sufficient depth to satisfy those with a keen interest in the history of science and technology. The book aims to not only inform but also inspire, demonstrating how the seemingly simple act of counting has been instrumental in shaping our understanding of the universe and transforming our world in profound ways. This journey is not simply about numbers; it is a reflection on human ingenuity and the enduring quest for knowledge.

Introduction

The seemingly simple act of counting underpins the vast tapestry of human achievement. From the earliest tally marks etched onto bone to the sophisticated algorithms driving modern quantum computers, counting has served as an indispensable tool for understanding and shaping our world. This book embarks on a journey to explore this fascinating evolution, tracing the development of counting from its rudimentary origins to its complex applications in contemporary science and technology. We will explore the early counting systems of ancient civilizations, Mesopotamia and Egypt, for instance, and examine how these systems facilitated the development of agriculture, commerce, and governance. We will see how counting enabled early astronomers to track celestial bodies, and navigators to chart courses across vast oceans and continents. The narrative then shifts towards the modern era, demonstrating the transformative impact of counting on the development of calculus, probability, statistics and, ultimately, the digital revolution. We will examine the mathematical underpinnings of quantum mechanics and artificial intelligence, highlighting the enduring relevance of counting in the twenty-first century. Throughout, the text will emphasize the interconnectedness of mathematical development with societal progress and cultural change. We will examine how the development of increasingly sophisticated counting systems both reflected and shaped the needs and aspirations of different societies. The book aims to offer a clear and engaging exploration of this crucial historical topic, making complex mathematical concepts accessible to a wide audience while remaining faithful to the nuanced insights of historical scholarship. This is a journey through human ingenuity, told through the lens of numbers.

Chapter 1: The Dawn of Counting: From Tally Marks to Tokens

Early Counting Systems: Evidence from Archaeology

The very origins of counting are shrouded in the mists of prehistory, yet the archaeological record offers tantalizing glimpses into the cognitive development of our ancestors. Long before the invention of writing or sophisticated number systems, humans devised ingenious methods for keeping track of quantities. These early counting systems, far from being mere curiosities, represent crucial steps in the evolution of human intelligence and social organization. Their study provides a fascinating window into the lives and minds of early humans, offering insights into their economic activities, social structures, and even their understanding of the natural world.

Perhaps the most compelling evidence for early counting comes from the discovery of tally marks etched onto bones and other artifacts. These marks, often simple scratches or incisions, represent the earliest known attempts at numerical representation. One of the most famous examples is the Ishango bone, discovered in the Ishango region of the Democratic Republic of Congo. Dating back to the Upper Paleolithic period (approximately 20,000 years ago), this bone fragment bears a series of notches arranged in distinct groups. The interpretation of these markings remains a subject of ongoing debate, with some scholars suggesting they represent a sophisticated lunar calendar or even a rudimentary understanding of prime numbers. While the precise meaning remains elusive, the Ishango bone undeniably demonstrates the capacity of early humans to conceptualize and represent numerical quantities.

The discovery of similar tally marks on bones from various archaeological sites across the globe supports the notion that the development of counting was not a localized phenomenon but rather a widespread human achievement. These marks, often found in association with hunting tools or other artifacts related to subsistence activities, suggest that early counting served practical purposes, such as keeping track of the number of animals hunted, the quantity of resources gathered, or perhaps even recording personal possessions. The simplicity of these systems, essentially a one-to-one correspondence between a mark and a unit, highlights the fundamental nature of counting: the ability to establish a quantitative relationship between an object and a symbolic representation.

The limitations of early tally mark systems are apparent. As the numbers being recorded grew larger, so too did the length and complexity of the tally. This inherent inefficiency likely prompted the development of more sophisticated counting systems. The next significant step in the evolution of counting is marked by the appearance of clay tokens in ancient Mesopotamia. Dating back to the late fourth millennium BCE, these tokens represent a profound advancement in numerical representation. Unlike simple tally marks, these tokens were three-dimensional objects, often shaped to represent the items they counted. For example, a cone-shaped token might represent a measure of grain, while a sphere might represent an animal.

The use of clay tokens represents a crucial transition from a purely unary system (one-to-one correspondence) to a more abstract form of numerical representation. By assigning a specific shape to represent a specific quantity or item, early Mesopotamians moved beyond the limitations of simple tally marks. However, the clay tokens themselves presented practical challenges. Managing a large number of tokens for complex transactions could become cumbersome, leading to the development of an ingenious system for recording token counts on clay tablets. This represents another pivotal moment in the history of counting, marking the beginnings of written numerical notations.

The tokens were impressed into the clay tablets, creating an imprint that served as a record of the quantity and type of items. This system, though still rudimentary, introduced the notion of symbolic representation, foreshadowing the development of more advanced writing systems. The discovery of these clay tablets offers invaluable insights into the economic and social activities of early Mesopotamian societies. They provide evidence for the existence of sophisticated trade networks, detailed accounting practices, and the organization of complex economic transactions. The clay tablets thus serve as a window into the practical application of early counting systems in the context of a rapidly developing civilization.

The transition from concrete representations of numbers (like tally marks and tokens) to more abstract symbolic representations (numerals) is a pivotal point in the history of mathematics. This shift wasn't merely a matter of convenience; it facilitated more sophisticated mathematical operations and abstract thought. It allowed for the development of arithmetic as we know it, enabling calculations that were previously impossible or incredibly laborious. Before numerals, computations were limited by the physical manipulation of objects or the length of tally marks. The abstract nature of numerals, however, freed mathematics from these physical constraints, allowing for the exploration of more complex mathematical relationships.

Consider the transition from Roman numerals to the Hindu-Arabic numeral system. Roman numerals, with their cumbersome representation of numbers using letters, were ill-suited for complex calculations. The Hindu-Arabic system, with its positional notation and the inclusion of zero, proved revolutionary. This system provided a more efficient way of representing numbers and performing calculations, paving the way for significant advances in mathematics, science, and technology. The adoption of this system marked a clear break from earlier, less flexible methods and demonstrated the power of symbolic representation in fostering mathematical innovation.

The archaeological evidence of early counting systems doesn't simply reveal the progression of numerical representation; it illuminates the inextricable link between mathematics and the human experience. The methods early humans used to count directly reflect their social and economic needs. Hunter-gatherer societies used simple tally marks to manage their resources and track their hunts, while the more complex societies of

Mesopotamia developed more sophisticated systems to handle the increased demands of agriculture, trade, and governance. This interconnection between counting and societal needs continued throughout history, shaping the evolution of mathematics and influencing the trajectory of human civilization. The archaeological record thus serves not only as a source of information about early counting but also as a evidence in the human ingenuity and adaptability that have driven the development of mathematics over millennia. It reveals a journey from simple tallies to complex systems, a journey that continues to this day.

The Development of Number Systems in Ancient Mesopotamia

The transition from clay tokens to a formalized system of numerical notation in Mesopotamia marks a pivotal moment in the history of mathematics. While the tokens themselves represented a significant advance over simple tally marks, their practical limitations—the sheer volume of tokens needed for complex transactions—spurred the innovation of cuneiform writing. This system, utilizing wedge-shaped marks impressed into wet clay tablets, allowed for a more efficient recording of numerical data. The development of cuneiform numerals wasn't a sudden leap but rather a gradual evolution, mirroring the complexities of the Mesopotamian society it served.

Initially, the cuneiform numerals directly mirrored the shapes of the clay tokens. A single vertical wedge might represent one unit of a particular commodity, while multiple wedges represented larger quantities. This system, though an improvement over managing individual tokens, still lacked the efficiency and flexibility of later number systems. The limitations become especially apparent when considering large numbers or complex calculations. Imagine the sheer number of wedges required to record a significant tax assessment or the inventory of a large-scale trading enterprise. The inherent clumsiness of this early cuneiform system would have quickly become apparent as Mesopotamian society grew more complex.

The breakthrough came with the adoption of a sexagesimal (base-60) system. Unlike our familiar decimal (base-10) system, the Mesopotamians developed a system using 60 as its base. The choice of 60 as a base is a subject of ongoing scholarly debate, with several plausible explanations offered. Some scholars suggest that it was influenced by the approximate number of days in a year (360) which is divisible by 60. Others point to its mathematical convenience, since 60 is divisible by many smaller numbers (2, 3, 4, 5, 6, 10, 12, 15, 20, and 30). This high divisibility makes calculations with fractions significantly easier than in base 10, allowing for more precise representation of quantities.

The Mesopotamian sexagesimal system was not positional in the same way as our modern base-10 system which uses a positional notation to differentiate between ones, tens, hundreds, and so on. For example, the number 123 is understood as 1 hundred, 2 tens, and 3 ones. While their system incorporated some positional elements, the placement of cuneiform symbols had more nuanced interpretations based on context. For instance, a specific wedge shape could signify 1, 60, 3600, or any other power of 60 based on its placement relative to other symbols. This subtle difference may not seem significant to modern eyes, but it reflects a different understanding of mathematics and demonstrates the remarkable ingenuity involved in creating a sophisticated system capable of handling complex computations.

The use of this system had a profound impact on Mesopotamian society. It facilitated the development of sophisticated accounting practices, essential for managing the complex agricultural economy. The fertile lands of Mesopotamia supported large-scale agriculture, requiring detailed record-keeping to track harvests, distribution of crops, and taxation. Cuneiform numerals allowed for efficient tracking of these vast quantities, enabling the administration of the complex agricultural economy that sustained their civilization.

Taxation was a cornerstone of Mesopotamian governance. The system enabled the state to efficiently levy taxes and collect revenues from its vast agricultural lands and trade networks. Clay tablets unearthed at numerous archaeological sites contain meticulously recorded tax assessments, providing invaluable insights into the economic workings of ancient Mesopotamian society. These tablets are not just dry records of numbers; they represent the intricate web of social and economic relationships that underpinned the Mesopotamian civilization. The analysis of these tablets has illuminated aspects of social stratification, land ownership, and the power dynamics within Mesopotamian society.

Beyond agriculture and taxation, Mesopotamian mathematics played a crucial role in the development of their extensive trade networks. The system allowed for accurate recording of trade transactions, inventory management, and the tracking of goods across vast distances. The efficiency of their numerical system was crucial for the success of their extensive trade networks that connected Mesopotamia with distant parts of the ancient world. Cuneiform tablets have been found across the region, demonstrating the remarkable reach of their commercial activities. The numerical precision of their records allowed them to track the flow of goods and manage their complex interactions with other civilizations.

Mesopotamian astronomers made significant contributions to the field, utilizing their sexagesimal system for celestial observations. Their astronomical calculations, often recorded on clay tablets, demonstrate a remarkable understanding of planetary movements and cycles. The precision required for astronomical calculations reinforced the usefulness of the sexagesimal system, highlighting its efficiency for representing fractions and dealing with complex mathematical relationships. The influence of Mesopotamian astronomy is far-reaching, with their observations and calculations influencing later astronomical traditions in Greece and other parts of the ancient world. The legacy of their astronomical achievements can still be seen in our modern usage of the 360-degree circle and the division of time into hours, minutes, and seconds.

The Mesopotamian sexagesimal system wasn't simply a tool for practical applications; it also fostered mathematical innovation. While their system wasn't fully positional in the modern sense, it laid the groundwork for the development of positional number systems in later civilizations. The influence of the Mesopotamian system can be observed in the development of Babylonian mathematics, which made significant contributions to algebra and geometry. Babylonian mathematicians developed sophisticated methods for solving quadratic equations and calculating areas and volumes of geometrical figures. Their achievements represent a high point of mathematical development in the ancient world.

The diffusion of Mesopotamian mathematical knowledge across ancient civilizations is a testament to its influence. Their numeral system, though not directly adopted in its entirety, influenced the development of mathematical practices in Egypt, Greece, and other parts of the ancient world. The exchange of mathematical ideas played a key role in the development of mathematics across different cultures. The Mesopotamian legacy is deeply embedded in the history of mathematics, showing how mathematical innovation is not isolated but rather a collaborative and evolving process across time and geography.

The evolution of the Mesopotamian number system, from simple tally marks to the complex sexagesimal system, represents a remarkable journey of human ingenuity. The system was not just a tool for counting; it was integral to the functioning of their society, supporting advancements in agriculture, trade, governance, and astronomy. The development of this sophisticated system stands as evidence of the deep interconnection between mathematical innovation and the societal needs of ancient civilizations, showcasing the role of mathematics in shaping the course of human history. The insights gleaned from the study of cuneiform tablets continue to illuminate the complexities of this ancient civilization and its significant contribution to the field of mathematics. The enduring legacy of their numerical system serves as a powerful reminder of the enduring importance of mathematics in the progress of human civilization.

Egyptian Hieroglyphic Numerals and their Applications

The transition from the practical, if cumbersome, token systems of Mesopotamia leads us naturally to the elegant, if ultimately less flexible, numerical system of ancient Egypt. Unlike the Mesopotamian sexagesimal system, the Egyptians employed a decimal (base-10) system, a choice likely influenced by the readily available ten fingers on human hands – a ubiquitous counting aid across cultures. This system, based on hieroglyphs, was remarkably consistent over millennia, offering a fascinating window into the mathematical practices of a civilization that flourished for over three thousand years.

The Egyptian hieroglyphic numerals are instantly recognizable, even to those unfamiliar with the complexities of the language itself. Rather than abstract symbols, the numerals were largely pictorial representations of quantities. The numeral for one, for instance, was a single vertical stroke, mirroring the simplest form of tally-keeping. Ten was represented by a horseshoe-shaped symbol, often interpreted as representing a hobble for cattle, a visually intuitive representation of a group. One hundred took the shape of a coiled rope, likely signifying the bundling of larger quantities for ease of transport and management. One thousand was depicted by a lotus flower, a powerful symbol in Egyptian culture and possibly reflecting the abundance associated with this number. Ten thousand was a finger pointing upward, possibly referencing the gesture used in counting. One hundred thousand was a tadpole or frog, while a million was depicted by a kneeling figure, perhaps signifying the vastness of such a quantity.

The beauty of the Egyptian system lay in its simplicity. Larger numbers were simply constructed by repeating the symbols. For example, the number 321 would be represented by three coiled ropes (300), two horseshoes (20), and one vertical stroke (1). This additive nature, while straightforward, also highlighted a key limitation: representing exceptionally large numbers required long sequences of symbols, leading to potential errors in transcription and calculation. However, for the scale of most transactions and projects in ancient Egypt, the system functioned remarkably well.

Archaeological evidence provides ample confirmation of the widespread use of hieroglyphic numerals. Inscriptions on tomb walls, temple carvings, and papyri all bear witness to their application. These inscriptions weren't mere decorations; they conveyed critical information, including the size of tributes offered to deities, the quantities of goods stored in granaries, and the measurements of monumental structures like pyramids. The sheer volume of these inscriptions is evident in the importance of accurate record-keeping in Egyptian society. The careful accounting of resources was essential for maintaining the complex administrative machinery that supported the pharaohs and the vast workforce engaged in monumental construction projects and agricultural activities.

Consider the construction of the pyramids, colossal monuments that stand as symbols of Egyptian power and engineering prowess. The precise coordination of thousands of workers, the transportation of enormous blocks of stone, and the meticulous placement of each stone all required a sophisticated system of measurement and quantification. Hieroglyphic numerals were essential to this process. Papyri unearthed near construction sites reveal detailed records of materials used, worker assignments, and even estimates of the amount of time required for various stages of construction. These documents, often written in cursive hieratic script, a faster, more abbreviated form of hieroglyphs used for everyday writing, illustrate the practical application of Egyptian numerals in one of history's most ambitious engineering feats.

Beyond construction, the Egyptian system played a vital role in managing their agricultural economy. The Nile River's annual flooding was crucial to Egyptian agriculture, yet its unpredictability necessitated careful planning and resource allocation. Hieroglyphic numerals enabled scribes to record harvest yields, tax assessments, and the distribution of grain. The accurate accounting of agricultural products was crucial for maintaining social order and preventing famine. Numerous papyri detailing tax records, grain inventories, and land surveys have been discovered, offering a detailed glimpse into the intricacies of the Egyptian agrarian system.

Astronomical observations also required precise numerical recording. Egyptians meticulously tracked the movements of celestial bodies, correlating them with the Nile's flooding and other significant events. Their calendar, based on a 365-day year, was remarkably accurate, showcasing their advanced understanding of celestial mechanics and their ability to represent astronomical data using their numerical system. While their astronomical understanding might not have matched that of the later Mesopotamians in sophistication, their consistent record-keeping, aided by their hieroglyphic numerals, provides invaluable insights into their observations and the development of their calendar system.

The Egyptian system, while practical for many purposes, did have limitations. The lack of a zero or place-value system meant that larger numbers required long strings of symbols. This made complex calculations more time-consuming and prone to errors, particularly when dealing with fractions. Unlike the Mesopotamian sexagesimal system, which facilitated the precise representation of fractions, the Egyptians employed a system of unit fractions, fractions with a numerator of 1, leading to cumbersome representations of more complex fractions. For example, representing 2/5 would be expressed as the sum of 1/3 + 1/15, rather than the simpler 0.4. This difference underscores the diverse approaches employed by ancient civilizations in developing mathematical tools suited to their specific societal needs.

Despite its limitations, the Egyptian hieroglyphic numeral system serves as a powerful demonstration of human ingenuity and the essential role of mathematics in societal organization. Its clear, intuitive nature, reflecting the simplicity of a base-10 structure, is compelling. Moreover, its sustained use for millennia shows its adaptability and effectiveness within the framework of ancient Egyptian society. The decipherment of hieroglyphic inscriptions, coupled with the archaeological context they offer, allows us to reconstruct a detailed picture of daily life, economic practices, and the grand engineering projects that shaped Egyptian civilization. The study of these numerals, thus, goes beyond simple numeration; it offers a fascinating insight into the cultural, societal, and intellectual fabric of one of the most influential civilizations in human history. The simplicity of the system, while perhaps limiting in comparison to later systems, ultimately reflects the practical needs and cultural context of its creation, reminding us that mathematical advancements are always shaped by the specific challenges and opportunities of their historical moment. The enduring legacy of the Egyptian hieroglyphic numerals stands as a powerful example of this interwoven relationship between mathematics, society, and human achievement.

Comparing Early Counting Systems Strengths and Weaknesses

The Egyptian system, with its straightforward decimal base and pictorial representations, provides a compelling contrast to the more abstract sexagesimal system of Mesopotamia. While the Mesopotamian system, with its base-60, allowed for greater precision in representing fractions, a crucial advantage in astronomical calculations and complex trade transactions, its reliance on cuneiform script and the inherent ambiguity of its symbol combinations presented challenges. The lack of a distinct symbol for zero, for example, required careful context to interpret numbers, increasing the potential for errors in transcription and calculation. Consider the difficulty of differentiating between 1 and 60, or 1 and 3600, without clear positional notation. This ambiguity, while overcome through careful conventions developed over centuries, highlights a fundamental difference in approach between the two civilizations.

The Egyptian system, by contrast, avoided this ambiguity through its reliance on simple repetition of symbols. However, this simplicity came at a cost. Representing large numbers required long strings of symbols, making calculations tedious and prone to errors, especially as the numbers increased. The lack of a zero also limited its ability to handle very large numbers or complex calculations efficiently. For instance, the multiplication of two large numbers would have involved a laborious process of repeated addition, in contrast to the potentially simpler algorithms facilitated by place value systems. Moreover, their approach to fractions, using unit fractions (fractions with a numerator of 1), led to complicated expressions for even simple fractions, further increasing the computational burden.

Moving beyond Mesopotamia and Egypt, we find further diversity in early counting systems. The Roman system, while familiar to many through its continued use in certain contexts, reveals a somewhat haphazard evolution. Its additive nature, with symbols representing specific values (I, V, X, L, C, D, M), mirrors the Egyptian system in its simplicity but lacks the consistency and elegance of a strictly decimal base. The subtractive principle, where a smaller symbol preceding a larger symbol indicates subtraction (IV = 4, IX = 9), introduced an element of complexity that was absent in the purely additive Egyptian system. This system, however, demonstrated flexibility in its use, adapting to varying scales of measurement and practical applications within the Roman Empire. Its widespread adoption across a vast and diverse empire underscores its effectiveness within a complex administrative and commercial context. Yet, its cumbersome nature for complex calculations and the lack of a place-value system hindered its capacity for substantial mathematical advancement.

Several other civilizations developed distinct counting systems reflecting unique cultural and environmental contexts. The Mayan civilization of Mesoamerica, for instance, employed a vigesimal (base-20) system, likely influenced by the twenty digits (fingers and toes) available for counting. This sophisticated system included a zero symbol, a crucial advancement that greatly facilitated calculations, and a place-value notation that enabled the representation of extremely large numbers. Their calendar system, based on this sophisticated numerical system, validates their mathematical understanding. The complexity of their astronomical observations and their ability to predict celestial events with astonishing accuracy underline the power of their numerical system. This system serves as a stark reminder that the seemingly arbitrary choice of a numerical base profoundly impacts the efficiency and scope of mathematical operations.

In contrast to the advanced Mayan system, some cultures relied on simpler tally-based systems. These systems, often employing notches on bones or marks on stones, were sufficient for tracking basic quantities and rudimentary transactions. While lacking the elegance and efficiency of later systems, they represent the foundational steps in the evolution of numerical representation. Archaeological discoveries of such tally marks, dating back to the Paleolithic era, provide direct evidence of early human efforts to quantify and organize their world. These systems, though seemingly primitive, are crucial for understanding the very genesis of counting and the cognitive development underlying our capacity for numerical abstraction.

The comparative analysis of these diverse systems reveals a fundamental truth about the development of mathematics: it is not a linear progression, but a complex interplay of cultural context, environmental pressures, and practical needs. The choice of a numerical base, the inclusion or exclusion of a zero, and the methods employed for representing fractions are not arbitrary decisions but reflect the specific challenges and opportunities faced by each civilization. The Mesopotamian preoccupation with astronomy, for instance, led to the development of a system particularly suited to fractional representation, while the practical demands of a large-scale agricultural society, such as that of Egypt, may have favored a system prioritizing simplicity and ease of use.

The strengths and weaknesses of each system, therefore, are not absolute but relative to their specific historical and cultural contexts. The Egyptian system, while simpler, lacked the precision and efficiency of the Mesopotamian system for certain tasks, yet it served the needs of Egyptian society remarkably well for millennia. Similarly, the Mayan system, with its sophisticated place-value notation and zero, surpassed both the Egyptian and Mesopotamian systems in its capacity for representing and manipulating large numbers. The Roman system, while less mathematically elegant, proved remarkably adaptable to the administrative needs of a vast empire.

Understanding the diversity and particularities of these early counting systems enriches our appreciation for the evolution of mathematics. It highlights the creative ingenuity of human societies in developing tools to quantify, measure, and understand their world, underscoring the crucial role of mathematics in the development of civilization itself. The transition from simple tally marks to the sophisticated systems of Mesopotamia, Egypt, and the Maya illustrates a progressive refinement of numerical representation, but also a dynamic interplay between mathematical innovation and societal needs. The story of these early counting systems is not merely a chronicle of mathematical progress but a rich narrative of human ingenuity and cultural adaptation. Each system, with its unique strengths and limitations, reflects a specific cultural moment, a demonstration of the power of human thought and the ongoing human quest to understand and quantify the world around us. The enduring legacy of these early systems lies not only in their historical significance but also in their contributions to the development of the more advanced numerical systems we use today.

The Transition from Concrete to Abstract Number: The Rise of Numerals

The shift from concrete representations of numbers to abstract numerals marks a crucial turning point in the history of mathematics. Early counting methods, relying on physical objects like tally marks etched into bone or pebbles arranged in patterns, directly mirrored the quantities being represented. A pile of five stones unequivocally represented the number five. This concrete approach, while effective for small numbers and simple transactions, presented inherent limitations. Scaling up to larger quantities became cumbersome, and complex calculations remained impractical. The need for more efficient methods of representing and manipulating numbers drove the development of abstract numeral systems.

This transition was a gradual process, unfolding differently across various cultures. Consider the Mesopotamian cuneiform system. While initially employing simple markings to represent quantities, the system gradually evolved into a more sophisticated sexagesimal (base-60) system. The use of cuneiform script allowed for the representation of larger numbers through combinations of wedge-shaped marks.

However, the absence of a distinct symbol for zero initially led to ambiguities. The same combination of marks could represent different values depending on context, necessitating careful interpretation. This ambiguity, while a significant limitation, also reflects the inherent challenges of representing abstract quantities using a limited set of symbols. The evolution of the system, including the eventual adoption of a placeholder symbol approximating zero, highlights the iterative nature of mathematical development. The system's inherent challenges spurred innovation and refinement, underscoring the relationship between limitations and progress.

The Egyptian hieroglyphic numeral system offers a different perspective. Its decimal base (base-10) provided a simpler structure, with distinct symbols for powers of ten (1, 10, 100, 1000, etc.). This system avoided the ambiguities of the early Mesopotamian system through its reliance on simple repetition. While straightforward, this additive nature presented practical limitations when dealing with large numbers. Representing a number like 999 required nine symbols for units, nine for tens, and nine for hundreds, making calculations laborious and prone to error. Moreover, the representation of fractions differed significantly from our modern approach, using unit fractions (fractions with a numerator of 1). For example, the fraction 2/3 would be represented as 1/2 + 1/6. This system, though less abstract than later systems, clearly moved beyond the purely concrete representations of earlier tallying methods. The inherent limitations of the Egyptian system, however, ultimately stimulated the development of more sophisticated computational methods and influenced the evolution of subsequent numeral systems.

The development of the Greek alphabetic numeral system demonstrates the evolving relationship between language and mathematics. This system, employing the Greek alphabet to represent numbers, provides an example of a system where abstract symbols were directly associated with numerical values. Each letter represented a specific number, and numbers were formed by combining these letter-symbols. However, this system, like the Roman system, lacked a consistent place-value notation, hindering its ability to efficiently represent and manipulate large numbers and complex fractions. Nonetheless, the alphabetic system illustrates the convergence of symbolic representation and numerical value, reflecting the integration of mathematical concepts within the framework of existing linguistic structures. The adoption of alphabetic numerals highlights the cultural context within which mathematical ideas were developed and disseminated.

The Roman numeral system, still familiar to many today, provides a fascinating case study in the limitations of a less abstract system. Its reliance on additive principles (combining symbols to represent larger numbers) and the subtractive principle (e.g., IV for 4, IX for 9) reflects a system that evolved pragmatically rather than following strictly mathematical principles. The Roman system's lack of a place-value system and the cumbersome representation of large numbers hindered mathematical progress within the Roman world. Complex calculations remained tedious, limiting the development of advanced mathematical techniques. The system's longevity, however, indicates its sufficient functionality for the administrative and commercial needs of a vast empire, further highlighting the interplay between mathematical sophistication and practical application.

In stark contrast to the Roman system, the Mayan civilization in Mesoamerica developed a remarkably advanced positional numeral system using a vigesimal (base-20) base. This system included a zero symbol and place value notation, which allowed for the concise representation of extremely large numbers. The Mayan system is evident of the potential for mathematical innovation, demonstrating the capacity for creating systems that effectively facilitated complex calculations and astronomical observations. Their precise calendar system and accurate astronomical predictions are direct consequences of this sophisticated numeral system. The Mayan system demonstrates the possibility of achieving a high degree of mathematical abstraction even in the absence of modern mathematical notation.

The evolution of numeral systems wasn't a uniform process. Different cultures adopted different approaches based on their specific needs and cultural contexts. Some cultures continued to rely on tallying systems, using simple marks to represent quantities, while others developed more complex systems. The diversity of these systems highlights the multifaceted nature of human cognition and the adaptability of mathematical thought. The development of abstract numeral systems involved not only mathematical innovation but also linguistic and cultural factors. The choice of symbols, the underlying base of the system, and the methods of representing fractions were all influenced by the specific cultural and historical contexts of the respective civilizations.

The transition from concrete to abstract number systems represents a fundamental shift in human cognitive ability. It involved the ability to detach numerical concepts from physical objects and manipulate them symbolically. This leap facilitated the development of more advanced mathematical operations and paved the way for abstract thought in other domains. The evolution of numerals showcases the interplay between practicality, ingenuity, and the development of abstract thought. The challenges and successes of each numeral system, from the simplest tally marks to the sophisticated Mayan system, provide a rich tapestry of human intellectual history. The story of the development of numerals illustrates the enduring human capacity for innovation and the remarkable resilience of mathematical thought across diverse cultures and historical periods. It is a story that continues to resonate in our modern world, built upon the foundations laid by these early systems of numerical representation. Understanding this historical evolution is crucial for fully appreciating the profound impact of numbers on human civilization. The seemingly simple act of counting has shaped our world in ways that extend far beyond the realm of mathematics itself, fundamentally influencing our scientific, technological, and societal advancements.

Chapter 2: Counting and the Development of Agriculture

Agricultural Accounting and Resource Management

The development of agriculture, a monumental shift in human history, was inextricably linked to the evolution of counting. While early hunter-gatherer societies had limited needs for sophisticated numerical systems, the transition to settled agricultural life brought about an unprecedented demand for tracking, managing, and distributing resources. The cultivation of crops, the herding of animals, and the storage of surpluses demanded a reliable method for quantifying and organizing these newfound assets. Archaeological evidence suggests that the emergence of agriculture directly spurred the development of more advanced counting techniques, moving beyond simple tallying to more structured systems capable of handling larger numbers and more complex calculations.

One of the earliest and most critical applications of counting in agricultural societies was inventory management. The successful storage of harvested crops was paramount to survival during lean seasons. Farmers needed a way to track the quantity of grain, fruits, vegetables, and other foodstuffs stored in granaries and other storage facilities. Initially, this likely involved simple tallying systems, using notches on bones or wooden sticks to represent each unit of a particular crop. However, as agricultural production increased and societies became more complex, these simple methods proved inadequate. The need for a more efficient and reliable system of record-keeping drove the development of more sophisticated methods.

Clay tokens, discovered at numerous archaeological sites across the Near East dating back to the Neolithic period, offer compelling evidence of early agricultural accounting. These tokens, often representing specific commodities like grains, sheep, or jars of oil, were used as physical representations of stored goods. The tokens themselves served as a direct representation of quantity. A pile of ten clay tokens depicting wheat clearly signified ten units of wheat. This system allowed for an effective inventory, offering a tangible representation of stored resources. This system was essentially a "concrete" numerical system, directly mirroring the quantities being counted. The act of counting was inextricably linked to the physical manipulation of the tokens themselves. As agricultural production and the complexity of storage increased, the need to manage larger quantities of tokens became apparent, leading to the eventual development of a more abstract system.

The evolution from clay tokens to more abstract numerical systems is a fascinating transition. As the number of tokens grew, managing them became increasingly difficult. The development of accounting systems using clay tablets to record the quantities represented by the tokens marked a significant step toward abstract representation. These tablets, imprinted with cuneiform script, provided a more durable and organized method for recording inventory. Instead of relying solely on the physical presence of the tokens, the tablets represented their quantities symbolically, thus initiating a movement away from the purely concrete and toward a more abstract form of numerical representation. This system was not merely a passive recording of existing quantities; it facilitated the organization and management of goods on a scale impossible with solely the use of tokens. This transition reflects a crucial stage in the development of agricultural accounting, demonstrating the interplay between practical necessities and mathematical innovation.

Beyond inventory management, counting was also essential for monitoring crop yields. Farmers needed a way to assess the productivity of their fields, comparing yields from year to year and making informed decisions about planting, irrigation, and other agricultural practices. This involved counting the number of harvested crops, and, in many cases, estimating the volume of grain or other goods produced. Early methods likely involved approximations, but the need for accuracy inevitably drove the development of more precise counting and estimation techniques. Evidence from ancient Egypt shows that even the earliest methods for assessing crop yields reflected a sophisticated understanding of quantitative relationships and the importance of measurement.

The distribution of agricultural surpluses was another area where counting played a vital role. In early agricultural societies, the redistribution of food and other resources was crucial for social cohesion and stability. Leaders needed a reliable way to allocate these resources among the members of their communities. This often involved the implementation of rationing systems, requiring the ability to count and divide quantities of goods among individuals or families. Evidence of this is found in ancient texts and archeological discoveries, often illustrating the methods used to allocate resources based on population size, social status, or other criteria. Such systems necessitate not only basic counting but also arithmetic operations, including division and subtraction, showcasing the rapid advancement of mathematical skills driven by the demands of agricultural management.

The link between counting and the rise of settled communities and civilizations is undeniable. The ability to manage resources effectively facilitated the growth of larger, more complex societies. The development of sophisticated agricultural accounting systems allowed for greater surplus production, supporting larger populations and leading to the emergence of specialized labor, social stratification, and ultimately, the foundations of civilization. The ability to track, manage, and redistribute resources efficiently was instrumental in fostering social and political stability. Agricultural surplus was not simply a matter of producing more food; it was a cornerstone of social organization, and the ability to manage that surplus efficiently depended fundamentally on the development of increasingly sophisticated systems of counting.

Moreover, the development of agricultural accounting was intrinsically linked to the development of writing systems. In many societies, the need to record agricultural data played a crucial role in the creation and evolution of writing. Early writing systems often started as simple accounting tools, evolving from pictographic representations of goods to more abstract symbols representing numerical values. The cuneiform script of Mesopotamia, for example, initially served as a system of accounting for agricultural products and other resources before developing into a more fully-fledged writing system. This connection highlights the deep interplay between the practical demands of agricultural management and the profound cognitive shift toward abstract symbolic representation.

Finally, it's important to consider the broader social and economic implications of agricultural accounting. The ability to track and manage resources empowered certain individuals or groups within society, leading to social hierarchies and power structures. Those skilled in accounting and resource management frequently held influential positions, reflecting the significance of these skills for the functioning of agricultural societies. The development of accurate agricultural accounting facilitated economic growth and trade, creating opportunities for the exchange of goods and services, further driving societal complexity and cultural exchange. The integration of accurate accounting systems with trade networks allowed for larger-scale transactions, further stimulating economic growth and influencing the development of more advanced mathematical practices. The historical record reveals a constant feedback loop between numerical innovation, economic development, and the strengthening of complex societies. The seemingly simple act of counting, therefore, had profound and wide-ranging consequences that extend far beyond the immediate sphere of agricultural production. It was a catalyst for social, economic, and intellectual advancement, shaping the development of civilizations in ways that remain deeply resonant today.

Land Measurement and Surveying

The efficient management of agricultural surpluses wasn't solely about storing grain; it also involved the crucial task of land management. The transition to settled agriculture necessitated a systematic approach to dividing, allocating, and utilizing land resources. This required a sophisticated understanding of measurement and geometry, a direct consequence of the increasing need to quantify and manage the physical space upon which agricultural production depended. Counting, therefore, expanded beyond simple inventory; it became the cornerstone of land surveying and the equitable distribution of arable land. The development of precise land measurement techniques was intrinsically linked to the evolution of agricultural societies, impacting social structures, infrastructure projects, and the very organization of civilizations.

In ancient Egypt, the annual flooding of the Nile River presented a unique challenge and simultaneously an impetus for advanced land surveying techniques. The fertile silt deposited by the Nile's inundation replenished the land, but the floodwaters also rearranged boundaries and obliterated previous markers. The precise re-establishment of property lines after each flood was essential for maintaining social order and preventing disputes over land ownership. This necessity fueled the development of sophisticated surveying techniques, requiring a detailed understanding of geometry and measurement. Egyptian surveyors, often referred to as "rope-stretchers," employed ingenious methods utilizing knotted ropes and geometrical principles to accurately demarcate land parcels.

These rope-stretchers utilized ropes marked with specific lengths to create right angles and measure distances. The process involved stretching ropes to form a right-angled triangle, often using a 3-4-5 ratio, ensuring accurate perpendicular lines. This simple yet effective technique allowed them to precisely measure and divide land, creating rectangular plots of consistent size. This was crucial for equitable distribution, fair taxation, and efficient agricultural planning. The development of standardized units of length, such as the cubit, further facilitated accuracy and consistency in land measurements. The cubit, based on the length of the forearm, was widely used across ancient Egypt and other civilizations, creating a standardized unit for comparing and recording land dimensions. The prevalence of standardized units highlights the crucial role of standardization in promoting trade, construction, and overall societal organization.

The precise nature of Egyptian land surveying is evidenced by the remarkable alignment and regularity of fields found throughout the Nile Valley. The careful division of land, far from being a mere practical necessity, had profound social and economic consequences. It contributed to the stability of the Egyptian state, by ensuring a predictable system of taxation and resource allocation. It also facilitated efficient irrigation systems, crucial for maximizing crop yields in a region largely dependent on the Nile. The systematic surveying of land, therefore, was not just a technical achievement; it was a fundamental component of the sophisticated organizational structures that characterized ancient Egyptian society.

Mesopotamia, another cradle of civilization, faced different, yet equally challenging, geographical conditions. The fertile crescent, characterized by its irregular terrain and the lack of regular flooding, necessitated a different approach to land surveying. Mesopotamian surveyors developed sophisticated techniques for measuring irregular land parcels, employing principles of geometry and trigonometry to accurately determine areas and boundaries. They developed methods to approximate the area of irregularly shaped fields, demonstrating a level of mathematical understanding well beyond basic arithmetic. Clay tablets containing geometrical problems and solutions, discovered at various Mesopotamian sites, highlight their advanced mathematical skills and their application to land surveying.

The development of cuneiform writing in Mesopotamia played a crucial role in documenting land surveys and property ownership. Clay tablets were used to record land boundaries, measurements, and ownership details, providing a permanent record that minimized disputes and facilitated the efficient transfer of property. These records not only preserved land ownership information but also provided valuable insights into the economic organization of Mesopotamian society. They illustrate the significant role land ownership played in the economic and social structures of the region, underscoring the close link between land surveying, record-keeping, and societal organization.

The impact of land measurement extended beyond agricultural land management. It became integral to large-scale construction projects such as temples, palaces, and irrigation canals. The precise alignment of structures and the creation of complex irrigation networks required accurate surveying techniques and a sophisticated understanding of geometry. The construction of monumental structures demonstrated the power of the state and served as symbols of its authority. The ability to accurately plan and execute these ambitious projects further solidified the position of rulers and underscored the significance of mathematical knowledge in maintaining social and political stability.

Furthermore, the development of land surveying spurred innovation in mathematics itself. The practical necessities of land measurement and division drove the development of geometrical concepts and techniques, leading to advancements in mathematics far beyond their immediate application to surveying. The need to calculate areas of irregular shapes, for instance, led to the development of sophisticated approximation methods and the refinement of geometrical principles. These advancements in mathematics had a ripple effect, impacting other areas of life, from architecture to astronomy. The close relationship between practical needs and theoretical advancements in mathematics is a recurring theme in the history of mathematical development.

The development of land surveying was not limited to ancient Egypt and Mesopotamia. Similar techniques were developed independently in other parts of the world, demonstrating the universality of the need for accurate land management in settled agricultural societies. Across various cultures, we see evidence of the evolution of standardized units of measurement, geometrical principles, and sophisticated surveying techniques, all driven by the practical necessities of managing land resources. These practices highlight the remarkable ability of human societies to adapt and innovate, demonstrating the ingenuity applied to address the challenges of resource management.

In conclusion, the development of land measurement and surveying was a critical component of the agricultural revolution. It wasn't merely a practical skill; it was a catalyst for social, economic, and mathematical advancements. The ability to accurately measure and divide land enabled the efficient management of resources, the equitable distribution of wealth, and the construction of large-scale infrastructure projects. The interplay between practical needs and mathematical innovation is clearly reflected in the evolution of surveying techniques, highlighting the deeply intertwined relationship between human ingenuity, societal organization, and the development of mathematical knowledge. The seemingly simple act of counting, therefore, evolved into a complex system of measurement and resource management, forming a vital foundation upon which complex civilizations were built. The legacy of these early surveying techniques continues to shape our understanding of land management and resource allocation today.

Calendars and the Measurement of Time

The meticulous record-keeping necessitated by agriculture's expansion naturally extended beyond the quantification of land and harvested goods. The cyclical nature of agriculture, the predictable rhythms of planting, growing, and harvesting, demanded a sophisticated understanding of time, leading to the development of calendars. These weren't merely tools for scheduling; they were fundamental to the organization and governance of early agricultural societies. The ability to predict seasonal changes, crucial for planting and harvesting, was a direct result of keen observation, rudimentary mathematics, and the growing capacity for abstract thought.

Early attempts at timekeeping likely involved simple observations of the sun's daily movement and the phases of the moon. The regular recurrence of daylight and darkness, coupled with the moon's cyclical transformations, provided the most readily available markers for tracking the passage of time. Rudimentary calendars, perhaps consisting of notched sticks or scratched stones, were used to record these cycles, reflecting the basic principle of tallying, a cornerstone of early counting systems. However, these early systems were inherently limited, lacking the precision needed for long-term agricultural planning or complex social organization.

The development of more sophisticated calendars required a deeper understanding of astronomical phenomena and a greater capacity for mathematical calculation. The observation of the stars and planets became increasingly important, as their regular movements provided a more accurate framework for timekeeping. Ancient civilizations meticulously charted the movements of celestial bodies, recognizing patterns and cycles that allowed them to predict seasonal changes with greater accuracy. This knowledge was not merely of scientific interest; it was profoundly practical, impacting every aspect of agricultural life. The ability to predict the onset of monsoon seasons, the flooding of rivers, or the commencement of planting seasons was vital for survival and prosperity.

The Sumerians of Mesopotamia, renowned for their advancements in mathematics and astronomy, created one of the earliest known sophisticated calendar systems. Their calendar, based on lunar cycles, consisted of 12 lunar months, each approximately 29.5 days long. However, the lunar year, totaling approximately 354 days, was shorter than the solar year (approximately 365 days). To reconcile this discrepancy, the Sumerians periodically inserted an intercalary month, a process that required careful astronomical observation and mathematical calculation. This adjustment underscores the inherent complexity of calendar construction, demanding a sophisticated understanding of both lunar and solar cycles.

The Egyptians, similarly dependent on the predictable flooding of the Nile, developed a calendar deeply intertwined with the river's annual inundation. Their calendar, based on the solar year, comprised 365 days, divided into twelve months of 30 days each, with five additional epagomenal days at the end. This calendar, remarkably accurate for its time, facilitated the organization of agricultural activities and the efficient management of resources. The accuracy of the Egyptian calendar allowed for the precise prediction of the Nile's flooding, a crucial event that determined the success of their agricultural cycle. The precision of their calendar is evident in the remarkable regularity of their monumental constructions, many of which were carefully aligned with astronomical events.

The Mayan civilization in Mesoamerica developed a calendar system of remarkable complexity, incorporating both solar and lunar cycles, as well as longer cycles spanning hundreds or even thousands of years. Their calendar comprised several interlocking cycles, including the Haab (a 365-day solar year) and the Tzolk'in (a 260-day ritual calendar). The precise interplay of these cycles created a complex system for tracking time, reflecting their sophisticated understanding of astronomy and mathematics. The intricate nature of the Mayan calendar highlights the cultural significance of timekeeping, extending beyond mere agricultural planning to encompass religious ceremonies, governance, and social organization.

The development of calendars across different cultures reveals a striking universality in the human need to organize time.

Despite variations in their specific systems, all calendars reflect a common objective: to structure and predict the passage of time in relation to naturally occurring cycles. The creation of calendars, however, involved more than just observation; it required a fundamental understanding of numerical relationships and patterns, advancing mathematical knowledge alongside the practical needs of agriculture.

The precise calculation of leap years, a challenge faced by many ancient calendar systems, showcases the advanced mathematical abilities of these societies. The need to reconcile the discrepancies between lunar and solar cycles, or to accurately predict the occurrence of eclipses, demanded a profound understanding of fractions, ratios, and mathematical algorithms. These calculations weren't merely theoretical exercises; they had direct implications for agricultural planning, religious observances, and social order.

The creation of calendars significantly impacted the structure of societies. Agricultural activities were scheduled around calendar dates, ensuring efficient utilization of resources and maximizing crop yields. Religious festivals and ceremonies were aligned with specific times of the year, reinforcing cultural values and promoting social cohesion. The calendar became a fundamental element of governance, providing a framework for regulating taxation, organizing labor, and maintaining social order. The ability to predict and manage time had a profound effect on the stability and prosperity of ancient civilizations.

Moreover, the development of calendars spurred further advancements in mathematics and astronomy. The need for accurate timekeeping fueled the development of sophisticated observation techniques, leading to a deeper understanding of celestial mechanics. The intricate calculations required for calendar construction, in turn, stimulated the development of mathematical algorithms and computational methods. The interplay between practical necessities and theoretical advancements highlights the dynamic relationship between mathematics, astronomy, and the organization of human societies.

The legacy of ancient calendar systems continues to shape our own understanding of time. Our modern Gregorian calendar, although refined over centuries, is rooted in the astronomical observations and mathematical principles established by ancient civilizations. The seemingly simple act of counting, fundamental to the agricultural revolution, evolved into a sophisticated system of timekeeping that profoundly influenced the social, religious, and political structures of early societies, creating a powerful example of the interwoven development of human knowledge, technology, and civilization. The history of calendars offers a compelling demonstration of the ingenuity of past societies and the enduring influence of their remarkable achievements.

The Impact of Agriculture on the Evolution of Counting Systems

The shift to agriculture, a pivotal moment in human history, wasn't merely a change in subsistence strategy; it was a catalyst for profound transformations in social organization, economic structures, and, significantly, the evolution of counting systems. Before the widespread adoption of agriculture, the need for numerical representation was relatively limited. Small, nomadic groups could manage their resources, a few tools, a small herd of animals, through simple one-to-one correspondences, perhaps using notches on bones or tally marks on stones. The demands of agriculture, however, rapidly escalated the complexity of counting. The cultivation of crops and the domestication of animals necessitated the tracking of larger quantities of goods, the management of landholdings, and the coordination of labor. This created a pressing need for more efficient and sophisticated methods of counting than simple tallying could provide.

The transition from scattered, nomadic existence to settled agricultural communities led to a dramatic increase in the scale of human endeavors. The cultivation of fields, the storage of harvested crops, and the management of livestock required a level of organizational complexity previously unknown. These activities demanded accurate record-keeping, efficient resource allocation, and precise measurements of land, produce, and time. Simple tally marks, while sufficient for small-scale needs, proved inadequate for the larger-scale demands of agriculture.

The need to track larger quantities of grains, livestock, and other resources spurred the development of more abstract counting systems. Early agricultural communities transitioned from concrete, one-to-one correspondences to the use of symbolic representations of numbers. This involved the creation of numerical systems with specific symbols representing different quantities. The Sumerians, for example, developed a sophisticated cuneiform writing system that included numerical notations. Their base-60 system, still visible in our measurements of time and angles, reflects the practical needs of managing large quantities of goods and the complexities of agricultural scheduling. The development of writing, closely intertwined with the evolution of counting, allowed for the recording of numerical data in a permanent and easily accessible format. This facilitated the transmission of knowledge across generations and enhanced the efficiency of administrative tasks in growing agricultural societies.

The management of land became a crucial aspect of agricultural life, necessitating the development of sophisticated surveying and land measurement techniques. Early agricultural societies needed to define property boundaries, allocate land parcels for cultivation, and estimate the size of fields. These tasks required a solid understanding of geometry and measurement, which in turn contributed to the refinement of counting systems. The use of standardized units of measurement, such as cubits or bushels, reflects a growing need for precision and accuracy in agricultural practices. The implementation of these standardized measurements required not only the ability to count but also to understand ratios and proportions, representing a significant leap in mathematical understanding.

The burgeoning economies of agricultural societies further drove the development of counting systems. Trade and commerce became increasingly important, requiring the ability to accurately assess the value of goods, conduct transactions, and record debts and credits. The emergence of marketplaces and the exchange of commodities necessitated the development of standardized units of currency and accurate methods of accounting. The need to track transactions, manage inventories, and assess profits stimulated the evolution of more sophisticated arithmetic operations, including addition, subtraction, multiplication, and division. These arithmetical advancements were deeply intertwined with the economic dynamics of agriculture-based societies, fostering a reciprocal relationship between mathematical progress and agricultural productivity.

The emergence of taxation, a critical aspect of early state formation, presented another significant challenge to counting systems. The need to assess and collect taxes from a large population required an efficient system for recording the amount of produce or livestock owed by each individual or household. This spurred the development of record-keeping systems capable of handling vast amounts of numerical data. Moreover, the ability to accurately calculate and distribute taxes demanded significant mathematical skills, contributing to the growth of expertise in counting and calculation. The development of taxation mechanisms is a clear example of how the practical needs of state administration further refined and advanced counting systems.

Beyond the immediate needs of agricultural production and economic transactions, the evolution of counting systems also impacted the development of other aspects of early civilizations. The construction of large-scale irrigation systems, monumental architecture, and public works projects required precise measurements, coordinated planning, and the efficient management of labor. These undertakings demanded a highly developed understanding of numbers and spatial relationships, contributing to the advancements in geometry, surveying, and engineering. The ability to count and measure accurately was crucial for the successful completion of these complex construction projects.

The growth of complex social hierarchies and administrative structures in agricultural societies also played a role in shaping the development of counting systems. The need to manage populations, allocate resources, and record genealogies pushed the boundaries of mathematical understanding. The organization of large-scale agricultural projects and the management of vast quantities of data required sophisticated counting and record-keeping systems. These developments are evident in the intricate administrative structures of ancient civilizations, which relied on accurate counting to maintain social order and manage resources effectively.

The reciprocal relationship between agriculture and the evolution of counting systems should not be underestimated. The demands of agriculture fueled the development of more sophisticated counting methods, while advancements in mathematics and numerical systems facilitated more efficient agricultural practices. This continuous feedback loop led to a complex interplay between human needs and intellectual progress, driving the development of both agricultural techniques and mathematical understanding. The evolution of counting, therefore, is not a linear progression but a dynamic process shaped by the complex interplay of social, economic, and environmental factors, with agriculture as a major driving force. The development of increasingly complex counting systems directly reflects the growing demands of increasingly complex agricultural societies. This relationship continues to resonate in our own world, where the sophisticated counting systems developed over millennia underpin the technological advancements that continue to transform our relationship with food production and distribution. The story of counting is, fundamentally, the story of human adaptation and ingenuity, with agriculture as its fertile ground.

Early Forms of Commerce and Trade

The burgeoning agricultural societies of the Near East and elsewhere didn't simply produce surpluses; they created the conditions for the emergence of organized trade and commerce. The very act of cultivating crops and raising livestock, creating a reliable and potentially abundant food supply, freed a portion of the population from the constant pursuit of sustenance. This allowed for specialization of labor, with some individuals focusing on crafts, manufacturing, or trading rather than solely on farming. This specialization, in turn, spurred the development of intricate networks of exchange, requiring increasingly sophisticated systems of counting to manage the complexities of trade.

Before the rise of agriculture, bartering was likely the primary mode of exchange. One-to-one trades, where goods were directly exchanged for other goods, were sufficient for the limited needs of small nomadic groups. However, the surplus production enabled by agriculture introduced a new dimension to exchange. The ability to accumulate goods beyond immediate needs created a demand for standardized units of value and more complex methods of accounting. The need to evaluate the relative value of different goods, a bushel of wheat versus a woven cloth, or a goat versus a pottery vessel, led to the development of rudimentary forms of currency and increasingly advanced numerical systems.

Early forms of currency were often based on readily available and valued commodities. In some regions, livestock served as a basic unit of exchange, with cattle or sheep being used to represent value. In others, grains, such as barley or wheat, became the preferred medium of exchange, their quantity carefully measured and counted. The importance of accurate counting is undeniable here; the success of a transaction hinged on the precise assessment of the quantity of goods exchanged. Without reliable counting, trade would be fraught with disputes and mistrust, hindering economic development.

The development of more abstract numerical systems was essential for managing the complexities of trade. Systems like the Sumerian sexagesimal system (base-60) with its cuneiform numerals, and the Egyptian hieroglyphic system, facilitated the recording of larger quantities and the performance of more complex calculations. These systems allowed merchants to track their inventory, manage their transactions, and calculate profits and losses. The use of standardized units of measurement for goods, weights, lengths, volumes, became crucial in ensuring fair and efficient trade. Without a common system of measurement, discrepancies in quantity could lead to disputes and hinder the smooth functioning of markets.

The emergence of long-distance trade routes dramatically increased the complexity of economic interactions. Ancient trade routes, such as the Silk Road connecting East and West Asia, the spice routes across the Indian Ocean, or the trans-Saharan trade, involved the exchange of a vast array of goods between geographically distant and culturally diverse populations. These long-distance trades demanded sophisticated systems for managing goods, tracking shipments, calculating exchange rates, and recording debts and credits. The distances involved often meant transactions spanned months or even years, requiring detailed records to ensure accountability and prevent disputes.

The impact of counting on long-distance trade wasn't limited to the mechanics of exchange; it also played a crucial role in shaping cultural interactions and economic development. The standardization of weights and measures across different regions, often facilitated by the adoption of common numerical systems, eased trade and fostered communication between diverse cultural groups. The successful management of long-distance trade required not only sophisticated mathematical skills but also extensive knowledge of geography, navigation, and diplomacy.

The significance of record-keeping in the development of commerce cannot be overstated. Clay tablets in Mesopotamia, papyrus scrolls in Egypt, and later, written records in diverse formats, served as vital tools for documenting transactions, tracking inventories, and managing accounts. These records provided evidence of economic activity, allowing historians to reconstruct the dynamics of ancient economies and the evolution of trade networks. The existence of such meticulous records highlights the importance that even early societies placed on precise accounting and transparent transactions. The development of sophisticated record-keeping systems was a vital prerequisite for the growth of complex economies and the emergence of large-scale trading networks.

The evolution of currency itself is inextricably linked to the development of counting systems. As societies transitioned from bartering to using more standardized mediums of exchange, the need for accurate counting intensified. The introduction of coined money, while a later development, represents a significant advancement in the sophistication of economic systems. Coined money provided a readily portable, divisible, and easily countable unit of value, simplifying transactions and making trade more efficient. The ability to count and record coins accurately became fundamental to the success of monetary economies.

The development of advanced counting systems and the associated mathematics of trade wasn't solely a matter of practicality; it also fostered innovation and facilitated economic growth. The ability to calculate profits, losses, and interest rates, all dependent on accurate counting, created new opportunities for investment and enterprise. The emergence of merchants as a significant economic force, with their specialized knowledge and skills in accounting and trade, testifies to the growing importance of mathematical capabilities in economic life. The refinement of counting systems and the associated mathematical advancements directly contributed to the expansion of trade and the growth of complex economies.

The story of early commerce is, therefore, fundamentally intertwined with the story of counting. From the simplest forms of bartering to the complexities of long-distance trade and the emergence of coined money, the ability to count accurately and efficiently was essential for every stage of economic development. The development of increasingly sophisticated numerical systems reflects the growing needs of increasingly complex economies. The legacy of these early systems—the impact of Babylonian mathematics on our units of time and angles, the Egyptian fractions, the persistence of certain counting practices in specific regions—demonstrates their lasting influence on our understanding and management of commerce and trade. The history of commerce is, in many ways, is evident in human ingenuity, adapting counting techniques to meet ever-expanding economic challenges. The story continues even today, as the complexities of global finance still rely on the fundamental principles of counting and calculation laid down in these earliest forms of commerce and trade.

Chapter 3: Counting in Astronomy and Navigation

Ancient Astronomical Observations and Calculations

The intricate relationship between counting and the burgeoning field of astronomy in ancient civilizations is validation of human intellectual curiosity and ingenuity. Long before the advent of sophisticated instruments, early astronomers relied on meticulous observation and increasingly sophisticated numerical systems to track the movements of celestial bodies, predict astronomical events, and develop calendars that governed agricultural practices and social life. The accuracy of their observations and calculations, often astonishing given their technological limitations, reveals a deep understanding of mathematical principles and a remarkable ability to apply these principles to the complexities of the cosmos.

The Babylonians, renowned for their advanced mathematical knowledge, made significant contributions to astronomy. Their sexagesimal number system (base-60), inherited from earlier Sumerian traditions, proved remarkably well-suited to astronomical calculations. The division of the circle into 360 degrees, still used today, is a direct legacy of this system. The Babylonians meticulously recorded astronomical observations over centuries, compiling detailed records of planetary positions, lunar cycles, and the occurrences of eclipses. These records allowed them to identify patterns and develop predictive models, demonstrating a sophisticated understanding of celestial mechanics. Their ability to predict eclipses, for instance, held immense cultural and religious significance, elevating their astronomers to positions of considerable influence within society. The intricate calculations underlying these predictions relied heavily on their advanced arithmetic and algebraic techniques, showcasing a remarkable integration of mathematical theory and observational data.

The development of sophisticated mathematical techniques wasn't simply a matter of theoretical exploration for the Babylonians; it was inextricably linked to their practical needs. Their calendar system, based on meticulous lunar observations, was essential for regulating agricultural activities and planning social events. Accurate prediction of seasonal changes was crucial for a society heavily reliant on agriculture, and the Babylonian astronomers, through their mastery of numerical calculation, provided this essential service. The precision of their astronomical calculations is also evident in their understanding of the precession of the equinoxes, a subtle shift in the Earth's axis that affects the timing of seasons. This understanding, although not fully understood in its totality, shows the depth of their observations and the accuracy of their mathematical models. Their work laid the groundwork for future astronomical advancements across various cultures.

The Egyptians, while employing a different numerical system, a decimal system based on hieroglyphs, also made significant contributions to astronomy. Their observations were less focused on predictive models of planetary movements than those of the Babylonians, instead, they concentrated heavily on the practical application of astronomy to their daily lives. The rising and setting of the sun and stars played a crucial role in their religious beliefs and agricultural practices, influencing the timing of religious ceremonies and planting cycles. Their knowledge of the celestial sphere was quite advanced; they constructed monumental structures, such as pyramids and temples, aligned with significant astronomical events like solstices and equinoxes. The precise alignment of these structures reflects an advanced understanding of geometry and astronomy, and highlights the importance of accurate measurement and calculation in their architectural achievements. The construction itself demanded a sophisticated understanding of surveying techniques and the ability to perform intricate calculations related to angles and distances.

The Egyptians' meticulous observations of the Nile's annual flooding, crucial to their agricultural economy, further demonstrates their practical approach to astronomy. They recognized the correlation between the flooding and the heliacal rising of Sirius, the brightest star in the night sky. The heliacal rising (the first appearance of a star in the morning sky after its period of invisibility) of Sirius coincided with the commencement of the Nile's inundation. This observation allowed them to develop a calendar system directly tied to the agricultural cycle. This sophisticated system, based on a 365-day year, was remarkably accurate, demonstrating their ability to carefully track and record celestial events over extended periods. This calendar system, despite its imperfections in comparison to later calendar systems, speaks volumes about the advanced capabilities in timekeeping and astronomical observation they possessed.

Beyond the Babylonians and Egyptians, other ancient cultures also contributed significantly to the development of astronomy. The Maya civilization in Mesoamerica, for example, developed a highly sophisticated calendar system incorporating complex cycles, and demonstrating an advanced understanding of celestial mechanics. Their calendar system, composed of interwoven cycles, provided a highly precise way to track time, incorporating both solar and lunar movements. The remarkable accuracy of their calendar calculations, far exceeding that of many of their contemporaries, is a striking example of how seemingly unrelated mathematical concepts can be combined for highly effective and practical results. Their astronomical observations were intimately linked to their religious beliefs, and their highly-skilled astronomers played important roles in both religious and political life.

The development of astronomical instruments further enhanced the accuracy and sophistication of ancient astronomical observations. While the earliest instruments were relatively rudimentary—simple gnomons (vertical sticks used to measure shadows and determine the sun's position) and crude sighting devices, their design and application relied upon a basic understanding of geometry and trigonometry. The improvement and refinement of these instruments over time reflects the increasing sophistication of mathematical knowledge. The improved accuracy of these instruments enabled astronomers to make more precise measurements and develop more accurate astronomical models. Examples include the armillary spheres used by both Greeks and Chinese for tracking celestial movements and determining their positions in the sky. The development and refinement of these instruments reflects the cyclical relationship between practical application and theoretical advancement; the need for improved accuracy often spurred the development of more advanced mathematical concepts.

The reciprocal relationship between astronomy and mathematics is undeniable. Astronomy provided a rich source of data that stimulated the development of new mathematical concepts and techniques. Conversely, advances in mathematics provided astronomers with the tools they needed to analyze their data, develop more accurate models, and make more precise predictions. The development of geometry, trigonometry, and arithmetic were all significantly influenced by the demands of astronomy, and, in turn, astronomy benefited from the ever-increasing mathematical tools that were available for its use. This illustrates the iterative, cyclical nature of knowledge development throughout history.

In conclusion, the story of ancient astronomical observations and calculations is a compelling narrative of human ingenuity and intellectual progress. Early astronomers, relying on their sophisticated counting systems and increasingly advanced mathematical techniques, were able to unlock the secrets of the cosmos, developing calendars, predicting astronomical events, and even influencing the design and construction of monumental structures. Their contributions represent a remarkable integration of observation, mathematical modeling, and practical application, laying the groundwork for the scientific revolution that would take place centuries later. The enduring legacy of their work, reflected in our modern systems of measurement and astronomical understanding, highlights the profound impact that these early civilizations have had on the development of both mathematics and astronomy as a scientific field. The meticulous records left behind by these ancient peoples offer not just a glimpse into their mathematical skills and their understanding of the cosmos, but also evident in the powerful human drive to understand and interpret the world around us.

The Mathematics of Navigation: Charting Courses Across Land and Sea

The intricate dance between counting and the exploration of both land and sea is a fascinating aspect of human history. Long before the advent of sophisticated instruments like sextants and compasses, navigators relied on a deep understanding of numerical systems, geometry, and keen observational skills to chart courses, estimate distances, and determine their locations. Their success depended not just on bravery and seamanship, but also on the application of mathematical principles that were as crucial as the wind in their sails or the stars in their skies.

Consider the remarkable navigational feats of the Polynesian peoples. Their voyages across the vast expanse of the Pacific Ocean, spanning thousands of miles, were not the result of blind chance, but rather a demonstration of their advanced knowledge of celestial navigation and sophisticated understanding of wave patterns, ocean currents, and the subtle cues offered by the natural world. Their counting systems, while differing in specifics from other cultures, allowed them to meticulously track their journeys, noting the passage of time, the distance traveled, and the direction of their course. This wasn't merely counting days; it involved correlating the positions of stars and constellations with the passage of time and the observed movement of their vessels.

Evidence suggests that Polynesian navigators developed highly refined techniques for estimating distances and directions. They observed the patterns of the swells, noting the direction and height of waves to gauge their proximity to landmasses or significant ocean currents. They utilized sophisticated methods for wayfinding, relying on a combination of observations of celestial bodies, the flight patterns of birds, and the changing colors and smells of the ocean. These methods were passed down through generations, often orally, and were intricately linked to their rich cultural heritage. While exact numerical systems varied across different Polynesian islands, a commonality was the careful observation and recording of navigational data, reflecting the importance of counting in their wayfinding practices. Their success in traversing such immense distances speaks volumes about their mathematical understanding and ingenious navigational techniques.

The development of sophisticated navigational charts, or star charts, further strengthened their abilities. These charts didn't depict landmasses in the familiar way we understand maps today; rather, they served as mnemonic devices, representing relationships between celestial bodies and the direction of islands and currents. This abstract method of charting required a deep understanding of geometry and the relationships between angles and distances, converting observational data into a practical navigational tool. The creation and interpretation of these charts involved counting and the application of spatial reasoning, showcasing a sophisticated level of mathematical competence. The navigational knowledge held by Polynesian navigators wasn't simply a practical skill, but also a complex body of knowledge integrated into their social and cultural structures. The elders, the holders of this vital knowledge, played a critical role in navigating, guiding the younger generations, and sustaining the cultural continuity of their voyages.

Moving eastward, across the vast continents of Asia, we find a different yet equally compelling narrative of navigation and counting. The Silk Road, a network of trade routes that linked the East and West for centuries, presented unique navigational challenges. Unlike seafaring, where the horizon and celestial bodies provided constant reference points, traversing the Silk Road meant navigating across varied terrains – mountains, deserts, and vast plains. Counting played a pivotal role in their success. Caravans maintained meticulous records of distances covered, supplies remaining, and the time spent on the journey. Distances were not measured in abstract units but in terms of days of travel, linking numerical counting with the practical reality of the journey. Landmarks, oases, and mountain passes served as navigational checkpoints, each meticulously recorded and passed down through generations of merchants and traders.

The development of rudimentary cartography also played a critical role in the success of Silk Road journeys. Unlike the abstract charts of the Polynesians, these maps were often more literal representations of the terrain, although the scale and level of detail varied significantly. Yet, even rudimentary maps demanded spatial reasoning and numerical precision. Estimating distances, marking the locations of water sources, and identifying key passages necessitated an understanding of basic geometry and a capacity for precise measurement, albeit with simpler tools than those used by seafaring cultures. Counting, however, was fundamental; it was the language through which the distances, the resources, and the crucial waypoints were translated into actionable knowledge. The survival and success of Silk Road caravans depended on careful planning, resource management, and accurate estimation, all reliant on the fundamental ability to count.

The interaction between different navigational systems and the exchange of navigational knowledge played a significant role in shaping the understanding and practice of navigation across different parts of the world. For instance, the contact between European explorers and Polynesian navigators often resulted in the adoption and adaptation of existing navigational techniques. While European technology, such as compasses and astrolabes, offered new tools for navigation, the inherent knowledge and experience of Polynesian navigators, particularly in the understanding of ocean currents and wave patterns, often supplemented and refined European methods.

The invention and development of the magnetic compass proved a revolutionary advancement in navigation. While its origins are uncertain and debated, its impact on seafaring was undeniable. Initially, the compass was a relatively simple tool, but its refinement and use in conjunction with other navigational techniques significantly increased the accuracy and efficiency of sea voyages. The mathematical principles underlying the compass's functionality were relatively straightforward, yet its impact was profound. The use of the compass reduced reliance on solely celestial navigation, allowing voyages to be undertaken in conditions of poor visibility. Its adoption represents a significant shift in navigational practice, emphasizing the continuous interplay between technological advancement and the application of mathematical principles. The development of improved compasses, more precise angular measurements, and the development of tools for celestial navigation further refined the ability to accurately plot courses, estimate distances, and determine locations.

Furthermore, the development of sophisticated cartographic techniques revolutionized navigation. The evolution from rudimentary sketches to detailed maps, reflecting more precise measurements and projections, relied heavily on the advances in mathematics, particularly geometry and trigonometry. The creation of accurate maps required not only a detailed understanding of the spatial relationships between different locations, but also the ability to translate these relationships into a two-dimensional representation. The mathematical knowledge required for cartography was complex and evolved over time, reflecting the ongoing interaction between mathematics and the practical demands of navigation. The rise of advanced cartography further expanded the ability of navigators to explore and chart previously uncharted territories, underscoring the interconnectedness of mathematics, technology, and human exploration.

In conclusion, the mathematics of navigation, whether across the open ocean or the vast expanse of land, represents a remarkable example of human ingenuity. From the sophisticated celestial navigation of Polynesian voyagers to the practical mathematics underpinning Silk Road caravans, and the revolutionary impact of the compass and advanced cartography, counting and geometry have been fundamental to human exploration and trade. The ability to count, to measure, and to extrapolate from observations has not only facilitated the exploration of the world but has also fueled the development of mathematical and technological advancements, creating a feedback loop between practical needs and theoretical innovation. The enduring legacy of these early navigational systems highlights the remarkable interplay between mathematical knowledge, technological innovation, and human exploration, as evident in our enduring curiosity and capacity for discovery.

Celestial Navigation and the Development of Trigonometry

The preceding discussion highlighted the crucial role of counting in navigation, emphasizing its practical applications in diverse cultures and geographical contexts. However, the evolution of navigation wasn't solely dependent on counting objects or measuring distances directly. The ability to accurately determine one's position at sea or on land also demanded a profound understanding of angles, triangles, and the relationships between them, a field of mathematics that ultimately blossomed into trigonometry. This intricate branch of mathematics proved instrumental in refining celestial navigation, enabling sailors and astronomers to calculate their location and time with significantly greater precision than ever before.

The seeds of trigonometry can be traced back to ancient civilizations, particularly those deeply involved in astronomy and agriculture. Early Mesopotamians and Egyptians, for example, possessed rudimentary forms of geometric knowledge, employing practical techniques to survey land, construct buildings, and predict astronomical events. They were aware of the relationships between the sides and angles of triangles, though their methods were largely empirical, lacking a formal theoretical framework. Their observations of celestial bodies, meticulously recorded over generations, inadvertently contributed to the emerging development of trigonometry. The observation of the apparent movement of the sun, moon, and stars across the sky, along with the seasonal changes they dictated, required a means of measuring angles and estimating distances, even if these calculations remained largely imprecise by modern standards.

The ancient Greeks, however, played a pivotal role in transforming empirical geometric knowledge into a more formalized and theoretical discipline. The works of Hipparchus of Nicaea (circa 190-120 BC) are particularly significant. Though his original writings have been lost, his influence on the development of trigonometry is undeniable, primarily through the work of later astronomers and mathematicians who referenced and expanded upon his findings. Hipparchus is credited with creating the first systematic trigonometric tables, which provided a means of calculating the lengths of sides and angles of triangles based on known values. This was a crucial development, marking a significant departure from solely empirical methods. His tables, likely based on observations of the celestial sphere, allowed astronomers to solve problems related to the positions of celestial bodies, predicting their movements with greater accuracy.

The development of the astrolabe, an ancient astronomical instrument, is closely intertwined with the advancement of trigonometry. The astrolabe, used since antiquity, enabled the measurement of celestial angles and the determination of time. Its construction and operation relied heavily on principles of spherical trigonometry, a subfield of trigonometry dealing with the relationships between angles and sides of spherical triangles. The astrolabe's ability to translate celestial observations into readily usable data was paramount in navigation and astronomy. While earlier versions of the astrolabe existed, the improvements and refinements made during the Greek era, influenced by the growing sophistication of trigonometric tables, enhanced its accuracy and functionality. The astrolabe became a critical tool for both astronomers and sailors, bridging the gap between theoretical calculations and practical applications.

The work of Claudius Ptolemy (circa 100-170 AD), an Alexandrian astronomer, geographer, and mathematician, represents a milestone in the history of trigonometry. His magnum opus, the *Almagest*, became the definitive astronomical text for over a millennium. Within the *Almagest*, Ptolemy meticulously developed and refined the trigonometric techniques pioneered by Hipparchus, including the creation of a more comprehensive table of chords, which served as a precursor to modern trigonometric functions like sine and cosine. Ptolemy's systematic approach to trigonometry, combined with his detailed astronomical observations, provided a powerful framework for solving a wide range of problems related to celestial positions and movements. His contributions weren't confined to theoretical advancements; his methods were directly applicable to practical navigation, enabling sailors to calculate their latitude and improve the accuracy of their course estimations.

The subsequent transmission and development of Greek trigonometry into the Islamic Golden Age (8th-13th centuries AD) further enriched the field. Muslim mathematicians, building upon the foundations laid by Greek scholars, made significant contributions to the theory and practice of trigonometry. They developed new techniques for solving trigonometric equations, introduced new trigonometric functions (like tangent and cotangent), and refined the existing trigonometric tables, increasing their accuracy and scope. Figures like al-Battani (c. 858-929 AD) and Nasir al-Din al-Tusi (1201-1274 AD) played crucial roles in these developments. Their work not only advanced theoretical understanding but also facilitated the creation of more sophisticated navigational instruments and improved the precision of astronomical calculations. Their contributions were integral to the advancement of celestial navigation, enabling long-distance sea voyages and the precise charting of territories.

The influence of Islamic trigonometry extended beyond the Islamic world, profoundly impacting the development of mathematics in Europe during the Renaissance. European mathematicians, such as Regiomontanus (1436-1476), actively engaged with and translated the works of Muslim scholars, integrating their advancements into the developing European mathematical tradition. Regiomontanus's *De triangulis omnimodis*, a comprehensive treatise on trigonometry, represents a pivotal moment in the transmission and transformation of mathematical knowledge. His work synthesized and expanded upon the achievements of both Greek and Islamic mathematicians, consolidating and systematizing trigonometry as a distinct branch of mathematics. This resulted in a further refinement of trigonometric tools and techniques, directly benefiting navigation, surveying, and astronomy. Regiomontanus's work helped establish trigonometry as a foundation for many scientific disciplines, significantly impacting the progress of navigation and exploration.

The subsequent centuries witnessed a continual refinement of trigonometric methods and the development of increasingly precise navigational instruments. The invention of the sextant, a crucial tool for measuring the altitude of celestial bodies, combined with improved clocks and astronomical tables, greatly enhanced the accuracy of celestial navigation. The ability to accurately measure the altitude of stars and the sun allowed sailors to determine their latitude with remarkable precision. This, along with increasingly detailed maps and charts, facilitated voyages of unprecedented length and scope, revolutionizing exploration and trade. The underlying mathematical principles driving these advancements, precisely the trigonometric calculations pioneered by Greek, Islamic, and later European mathematicians, were instrumental in this transformation.

In conclusion, the development of trigonometry wasn't a singular event but a gradual evolution spanning centuries and numerous cultures. From its rudimentary beginnings in the empirical geometry of ancient civilizations to the sophisticated theoretical framework developed by Greek, Islamic, and European mathematicians, trigonometry profoundly impacted the field of celestial navigation. The ability to precisely calculate angles, distances, and positions enabled sailors and astronomers to chart courses with greater accuracy, facilitating exploration, trade, and a deeper understanding of the cosmos. The interwoven history of counting, geometry, and trigonometry underscores the profound interconnectedness between mathematical innovation and the advancement of human civilization. The legacy of these intertwined fields continues to shape our understanding of the world and our place within it.

Mapping and Cartography Representing the World Through Numbers

The preceding discussion illuminated the vital role of numerical systems in celestial navigation, a field that relied heavily on the precise measurement of angles and distances. However, the application of counting extended far beyond the realm of the stars; it played an equally crucial role in the development of another crucial tool for exploration and understanding our world: maps. The history of cartography, the art and science of mapmaking, is inextricably linked to the evolution of counting and mathematical techniques. Early maps, while rudimentary compared to modern cartographic products, still relied fundamentally on numerical data, even if this data was often approximate or based on estimations.

The earliest forms of cartographic representation were far from the precise and detailed maps we are familiar with today. Instead, they were often simple sketches, diagrams, or itineraries that conveyed general spatial relationships rather than accurate geographical information. Cave paintings, for example, sometimes depict the layout of a landscape, perhaps showcasing the location of water sources, hunting grounds, or significant landmarks. While not maps in the strictest sense, these early representations laid the groundwork for later, more sophisticated cartographic endeavors. These early attempts to represent spatial information, however rudimentary, demonstrate an inherent human need to understand and communicate the layout of their surroundings. The inclusion of even basic numerical information, such as the estimated distance to a specific location, would have added to the practical value of such representations.

Ancient civilizations, particularly those with established agricultural practices and complex societal structures, gradually moved beyond simple sketches and towards more formalized mapmaking. The need to delineate property boundaries, manage irrigation systems, and plan urban development created a demand for more accurate and reliable spatial representations. The development of standardized measurement systems, closely tied to the advancement of counting techniques, greatly enhanced the ability to create more precise maps. The ancient Egyptians, for instance, employed a sophisticated system of land surveying, known as *geodesy*, to delineate boundaries and assess land areas for taxation purposes. Their precise methods, coupled with their understanding of geometry, enabled them to create reasonably accurate maps for managing their agricultural lands and building projects. These maps, while not using the same mathematical projections as modern maps, still required careful measurement and the application of numerical data to represent the relative positions of geographical features. The reliance on standardized units of measurement was a crucial step in the evolution of cartography.

The Babylonian civilization also made significant contributions to cartography. Their extensive knowledge of astronomy and their proficiency in mathematics facilitated the creation of maps that incorporated celestial information. Their maps often included celestial coordinates, alongside terrestrial features. This integration of celestial and terrestrial data improved the accuracy of navigational maps, crucial for long-distance trade and military campaigns. The Babylonian sexagesimal (base-60) numeral system, with its ability to represent fractions with remarkable accuracy, played a key role in these advancements. The precision afforded by the sexagesimal system allowed for more refined measurements of angles and distances, leading to more accurate representations on their maps.

The ancient Greeks made substantial contributions to both mathematics and cartography, advancing mapmaking techniques considerably. They recognized the spherical nature of the Earth, a crucial step in the development of more accurate world maps. Anaximander (610-546 BC), a pre-Socratic philosopher, is credited with creating one of the earliest known world maps, depicting the Earth as a flat disc surrounded by water. While this representation was inaccurate in its depiction of the Earth's shape, it represented a significant step towards a more systematic approach to mapmaking. Subsequent Greek cartographers, such as Hecataeus of Miletus (c. 550-476 BC) and Ptolemy (c. 100-170 AD), built upon earlier work, incorporating more geographical information and refining their mapping techniques. Ptolemy's *Geographia*, a comprehensive atlas featuring maps of the known world, became a standard reference for centuries. Its influence on later mapmaking was profound, demonstrating the lasting impact of numerical data and geometrical principles on cartographic accuracy. Ptolemy's work relied heavily on a grid system for locating places using latitude and longitude, concepts that deeply relied on precise numerical calculation and a thorough understanding of geometry and trigonometry.

The Roman Empire also produced impressive maps, many of which were designed for practical purposes, such as military campaigns and administrative governance. The Romans were masters of surveying and engineering, and they employed sophisticated techniques to map their vast empire. Their road maps, meticulously documented using standardized units of measurement, facilitated communication and transportation throughout the empire. These maps often included numerical data representing distances between locations, allowing travelers to plan their journeys effectively. The sheer scale of the Roman Empire necessitated highly accurate mapping for both military operations and efficient administrative control. Their contributions represent a significant advancement in the practical application of mapping using precise numerical data.

The development of projection techniques marked a significant advance in the history of mapmaking. Since the Earth is a sphere, representing its surface accurately on a flat map presents a challenge. Different projection techniques were developed to address this challenge, each with its own strengths and weaknesses. The choice of projection depended on the map's intended purpose, its scale, and the specific region being mapped. Each projection involved complex mathematical calculations using geometrical principles and trigonometry, all deeply rooted in counting and numerical analysis. The development of different projection systems highlighted the increasing sophistication of mathematical tools used in cartography, directly reliant on numerical methods for calculation and precision.

During the Age of Exploration (15th-17th centuries), advancements in navigation and cartography went hand-in-hand. The improved accuracy of astronomical instruments, such as the astrolabe and the sextant, combined with more advanced mathematical techniques, allowed explorers to chart new lands with greater precision. The use of latitude and longitude, expressed numerically, became increasingly crucial in locating positions on the globe. The data collected by explorers, often meticulously documented through counting and measurement, served as the foundation for more accurate and comprehensive maps. The resulting maps were essential for navigation, trade, and colonization, shaping the world's political and economic landscape.

The subsequent centuries witnessed a continuous refinement of cartographic techniques and the increasing use of sophisticated mathematical tools. The development of geodesy as a scientific discipline, coupled with advancements in surveying techniques, significantly enhanced map accuracy. The advent of photography and later satellite imagery revolutionized mapmaking, providing unprecedented levels of detail and precision. However, even in the age of digital mapping, the fundamental principles of counting, measurement, and mathematical calculations remain at the core of cartographic practice. Modern Geographic Information Systems (GIS) rely heavily on numerical data to store, analyze, and visualize spatial information. The underlying mathematical frameworks, deeply intertwined with counting, are instrumental in the accurate representation of geographical features and the analysis of spatial data.

In conclusion, the relationship between counting and mapmaking is a long and deeply intertwined history. From rudimentary sketches to sophisticated digital maps, the ability to represent the world through numbers has been fundamental to the development of cartography. The evolution of mapmaking techniques reflects the progress of mathematical knowledge, from basic measurement systems to advanced geometrical projections and computer algorithms. The ongoing integration of mathematical and computational methods in cartography ensures that its capacity to represent and analyze spatial information continues to evolve. The history of maps is therefore not just a history of geographical representation, but also a history of numerical innovation and its application to the spatial understanding of our world.

The Influence of Astronomy and Navigation on Mathematical Development

The seemingly boundless expanse of the night sky, speckled with countless stars, has captivated humanity since time immemorial. Early civilizations, far from being passive observers, actively engaged with the heavens, developing sophisticated systems for tracking celestial movements. This engagement, driven by practical needs such as agriculture, timekeeping, and navigation, profoundly impacted the development of mathematics. The precise observation and prediction of celestial events necessitated the creation of ever more refined mathematical tools, fostering a dynamic interplay between theoretical advancements and practical applications.

One of the earliest examples of this interplay can be seen in the development of sophisticated calendars. Precise agricultural practices demanded an accurate understanding of seasonal cycles, closely tied to the movements of the sun, moon, and stars. Ancient civilizations, such as the Egyptians and Babylonians, developed complex calendrical systems based on meticulous observations of celestial bodies. These systems required a deep understanding of numerical relationships, leading to advancements in arithmetic and the development of more refined methods for counting and calculating. The Egyptian calendar, for instance, was remarkably accurate, reflecting a sophisticated grasp of astronomical cycles and a refined understanding of numerical relationships. Their ability to predict the annual flooding of the Nile, crucial for their agricultural success, was directly linked to their astronomical observations and mathematical calculations. The Babylonian calendar, based on a lunisolar system, was equally impressive, showcasing their advanced knowledge of celestial mechanics and their skills in mathematical computation.

The need for accurate timekeeping further fueled mathematical innovation. The construction of sundials, water clocks, and other timekeeping devices demanded an understanding of geometry and the consistent application of measurement. Sundials, in particular, relied on the precise measurement of angles and the careful calculation of shadow lengths, leading to advancements in geometry and trigonometry. The accuracy of these devices was directly dependent on the mathematical precision employed in their design and construction. The development of more accurate timekeeping mechanisms drove the refinement of mathematical techniques, resulting in better tools for understanding and predicting celestial events.

Navigation, perhaps even more than agriculture or timekeeping, exerted a powerful influence on the development of mathematics. Long-distance seafaring, a hallmark of many ancient cultures, required a sophisticated understanding of celestial navigation. The ability to determine one's location at sea, essential for successful voyages, hinged on the precise observation and measurement of celestial objects. Early navigators relied on simple techniques, such as observing the position of the sun and stars, but as voyages grew longer and more ambitious, the need for more precise methods became increasingly apparent. This demand directly stimulated the development of advanced mathematical tools.

The invention of the astrolabe, a crucial navigational instrument, demonstrates the symbiotic relationship between astronomy and mathematics. This ingenious device, capable of measuring the altitude of celestial bodies, combined a deep understanding of spherical trigonometry with precise craftsmanship. Its development required significant advances in geometrical and trigonometric calculations, facilitating more accurate determination of latitude at sea. The astrolabe's effective use relied on an intricate understanding of spherical geometry, allowing navigators to solve complex problems of celestial positioning and terrestrial location. This spurred further advancements in trigonometry, particularly in the development of trigonometric tables that aided in quick calculations during voyages.

The development of trigonometry, in particular, is inextricably linked to the needs of astronomy and navigation. The calculation of angles and distances in both celestial and terrestrial contexts necessitated the creation of trigonometric functions, such as sine, cosine, and tangent. These functions, essential for solving geometrical problems related to celestial observations and terrestrial surveying, played a crucial role in improving the accuracy of navigation. The construction of trigonometric tables, a monumental task that required immense computational skill and a deep understanding of mathematical principles, stands as a significant achievement in the history of mathematics. These tables simplified the calculations required for navigation and astronomy, making them more accessible to a wider range of practitioners.

The interaction between theoretical mathematics and its practical applications in navigation and astronomy was not a one-way street. Advances in pure mathematics often found unexpected applications in these fields, leading to further refinements and improvements. For example, the development of spherical trigonometry, driven by the needs of astronomy, had far-reaching implications for cartography and surveying. The ability to accurately represent the curved surface of the Earth on a flat map, a complex geometrical problem, relied heavily on the principles of spherical trigonometry. Similarly, the development of algebra provided a powerful tool for solving complex astronomical and navigational calculations, leading to more precise predictions and measurements.

The work of Ptolemy (c. 100-170 AD) exemplifies this powerful interplay between theoretical and applied mathematics. His *Almagest*, a comprehensive treatise on astronomy, provided a detailed mathematical model of the solar system, incorporating sophisticated geometrical calculations. This model, although ultimately superseded by later, more accurate models, represented a remarkable achievement in the application of mathematics to the understanding of celestial phenomena. Ptolemy's work, extensively using trigonometry and advanced geometrical methods, became a standard text for centuries, showcasing the enduring influence of the integration of mathematics and astronomy. His work not only profoundly impacted astronomy but also laid the groundwork for subsequent advancements in mathematics. His contributions to cartography, specifically his *Geographia*, similarly demonstrated the power of applying mathematical principles to solve practical problems, particularly the problem of mapping a spherical surface onto a flat plane.

The Age of Exploration (15th-17th centuries) witnessed another significant surge in the development of mathematics driven by the demands of navigation. The expansion of maritime trade and exploration necessitated more accurate navigational techniques. The development of improved instruments, such as the sextant and the mariner's compass, along with advancements in cartography and surveying, required increasingly sophisticated mathematical tools. The ability to accurately determine longitude, a long-standing challenge for navigators, was ultimately solved through the development of precise chronometers, devices that required intricate mechanical engineering and accurate mathematical modeling of timekeeping mechanisms. The solution to the longitude problem highlights the close relationship between mathematical advancement and technological innovation spurred by practical necessities.

The development of calculus, a major turning point in the history of mathematics, also owes a debt to the study of astronomy and celestial mechanics. Isaac Newton's work on gravitation, a landmark achievement in the understanding of planetary motion, relied heavily on the tools of calculus. The ability to accurately model the movements of celestial bodies, a central concern of astronomy, provided a powerful impetus for the development of calculus, a branch of mathematics that has subsequently found countless applications in diverse fields.

In conclusion, the relationship between astronomy, navigation, and mathematical development is one of profound interdependence. The practical needs of these fields, specifically the demands for accurate timekeeping, navigation, and celestial observation, spurred the creation of new mathematical tools and techniques. This close interplay between theoretical advancements and practical applications fueled mathematical innovation, enriching both our understanding of the cosmos and our ability to navigate and explore our world. The history of mathematics is not just a history of abstract thought but also a history of practical problem-solving, intimately connected to humanity's ongoing endeavor to comprehend and master its environment. The seemingly distant stars, once merely points of light, became, through meticulous observation and the application of ever more sophisticated mathematical tools, crucial components in the development of human knowledge and civilization.

Chapter 4: Counting and the Rise of Cities

Urban Planning and Infrastructure

The intricate dance between the heavens and human ingenuity, as explored in the preceding sections, finds a striking parallel in the rise of cities and the development of urban planning. Just as the need for accurate navigation and astronomical prediction fueled mathematical innovation, so too did the demands of creating and managing complex urban environments. The transition from scattered settlements to densely populated cities demanded sophisticated systems for organizing space, distributing resources, and coordinating human activity. This necessitated a profound understanding of numbers and the application of geometrical principles, demonstrating once again the inseparable link between mathematical advancement and societal progress.

The very act of establishing a city involved a fundamental engagement with numbers. Determining the optimal size and layout of a city, considering factors like population density, resource availability, and defensive needs, required meticulous planning and a sophisticated understanding of spatial relationships. Early city planners, though lacking the tools of modern engineering, demonstrated an intuitive grasp of geometric principles. The use of regular grids, as evidenced in cities like Mohenjo-daro in the Indus Valley Civilization, suggests a conscious effort to create efficient and easily navigable urban spaces. The precise alignment of streets, buildings, and other structures indicates a level of mathematical sophistication that belies the apparent simplicity of the designs.

The construction of monumental architecture, a defining feature of many ancient cities, demanded even greater mathematical proficiency. The construction of pyramids in ancient Egypt, ziggurats in Mesopotamia, or the great temples of the classical world, involved intricate calculations concerning angles, dimensions, and materials. These monumental projects were not simply feats of engineering prowess, but also confirmations of the mathematical knowledge and organizational capabilities of their builders. The precise alignment of the pyramids, for example, demonstrates an astonishingly accurate understanding of geometry and surveying techniques. The massive scale of these structures necessitates an understanding of ratios, proportions, and the accurate calculation of volumes. The use of sophisticated tools and techniques for measuring angles, lengths, and volumes attests to the advanced mathematical skillset necessary for their construction. These projects not only showcase architectural innovation, but serve as evidence of the concurrent advancements in mathematics.

The distribution of resources within a city also depended on numerical expertise. The management of water supplies, a crucial aspect of urban planning, often involved complex irrigation systems, canals, and aqueducts. Ancient Mesopotamia, renowned for its sophisticated irrigation networks, relied heavily on mathematics to design and manage its water resources. The precise calculation of water flow, gradient, and canal dimensions required a deep understanding of arithmetic and geometry. The efficient management of water played a pivotal role in the success of Mesopotamian cities and reflected their mastery of quantitative analysis. Likewise, the distribution of food, building materials, and other essential goods necessitated effective inventory management and logistics, tasks that required numerical tracking and analysis. These systems, though often rudimentary by modern standards, demonstrate the practical applications of counting and calculation in everyday urban life.

The Roman Empire provides a compelling case study of the interplay between urban planning and mathematics. Roman cities were renowned for their efficient layout, based on a grid system that facilitated navigation and urban management. The precise alignment of streets, the consistent dimensions of buildings, and the construction of extensive infrastructure, such as aqueducts and roads, all reflect a deep understanding of geometry and surveying techniques. The Roman road network, for instance, is evident of the Roman engineering prowess and their mathematical expertise. The precise measurements and careful construction of Roman roads facilitated trade, communication, and the efficient movement of troops across vast distances. The Roman's adoption of standardized measures of length and area enabled the efficient building and maintenance of this vast infrastructure.

The construction of Roman aqueducts, renowned for their ability to transport water over long distances, demonstrates a sophisticated understanding of hydraulics and engineering principles. The successful design and construction of aqueducts required complex calculations of water pressure, flow rate, and the gradient of the aqueducts. This necessitated a deep understanding of geometrical concepts as well as practical knowledge of material properties and construction techniques. Moreover, the systematic surveying and precise measurements crucial for the construction of aqueducts were enabled by an advanced understanding of geometry and trigonometry. The meticulous attention to detail evident in Roman engineering projects highlights the importance of mathematical proficiency in urban planning and infrastructure development.

Beyond the physical infrastructure, the administration of Roman cities required sophisticated numerical systems. Tax collection, census taking, and the management of public finances all necessitated the use of counting and calculation. The Roman numeral system, although less efficient than positional systems, played a vital role in the administration of the vast Roman empire. The development of sophisticated accounting methods allowed the Romans to efficiently manage their finances and track their resources. The accurate record-keeping essential for efficient governance required a considerable understanding of numerical operations.

The examples of Rome and Mesopotamia are not isolated instances. Throughout history, the development of cities has been closely intertwined with mathematical progress. From the layout of streets in ancient Indus Valley cities to the construction of monumental structures in pre-Columbian America, the application of mathematical principles has been essential in creating functional and sustainable urban environments. The challenges of urban planning, managing population density, distributing resources, and coordinating complex social interactions, have continuously pushed the boundaries of mathematical knowledge and driven the development of new tools and techniques. The history of cities, therefore, is not simply a history of architecture and engineering but also a history of the application of mathematics to the solution of practical problems, shaping not only the physical landscape of urban spaces but also the social and economic fabric of human societies. The seemingly mundane act of counting, therefore, holds a profound significance in the unfolding narrative of human civilization, its influence extending far beyond the realm of pure mathematics to shape the very environments in which we live. The rise of cities, then, was not only a demonstration of human ingenuity and organization but also a direct reflection of the developing sophistication of mathematical understanding. The ability to quantify, measure, and predict became increasingly crucial to the success and sustainability of these burgeoning urban centers, creating a dynamic feedback loop where the demands of urban life spurred further mathematical innovation, culminating in the complex and technologically advanced urban landscapes we inhabit today.

Taxation and Resource Allocation

The seemingly simple act of counting took on a profound significance in the burgeoning cities of antiquity, extending far beyond mere enumeration to become the bedrock of governance and societal stability. Taxation and resource allocation, crucial elements in maintaining social order and funding public works, were inextricably linked to the development and refinement of numerical systems. The effectiveness of these systems, in turn, directly influenced the prosperity and longevity of these early urban centers. The transition from small, relatively self-sufficient settlements to complex, densely populated cities necessitated sophisticated mechanisms for collecting revenue, distributing goods, and managing resources, all of which relied heavily on the ability to accurately count and record numerical data.

One of the earliest examples of the application of counting to taxation can be found in ancient Mesopotamia. The Sumerians, renowned for their advanced civilization, developed a sophisticated system of record-keeping, utilizing cuneiform script to document economic transactions. Clay tablets, unearthed by archaeologists, reveal detailed records of agricultural yields, livestock counts, and tax assessments. These tablets demonstrate that taxation was not a haphazard process but rather a carefully planned and meticulously documented system. Taxes, often levied as a percentage of agricultural output or as a portion of livestock holdings, provided the revenue necessary to fund public works, including irrigation systems, temples, and palaces. The accuracy of these records was crucial, not only for ensuring equitable taxation but also for maintaining the economic stability of the city. Errors in counting could have significant consequences, potentially leading to social unrest or economic hardship.

The sophistication of the Mesopotamian tax system is further evidenced by the development of specialized administrative roles. Scribes, individuals trained in the art of writing and mathematics, were entrusted with the task of maintaining meticulous records of tax assessments and revenue collection. Their proficiency in numeracy was essential for the smooth functioning of the Mesopotamian economy and governance. The use of clay tablets, though seemingly rudimentary, allowed for the creation of permanent records, minimizing the potential for corruption or manipulation of tax data. Furthermore, the development of standardized weights and measures facilitated accurate assessment of taxable goods, adding further precision to the tax system. These standardized units, carefully calibrated and widely accepted, ensured fair and consistent taxation across the various city-states of Mesopotamia.

The Egyptian civilization also employed sophisticated counting systems for taxation and resource management. The Nile River, the lifeblood of Egypt, necessitated a highly organized system for managing water resources and distributing land. The pharaoh, as the supreme ruler, controlled the land and oversaw the allocation of resources, including agricultural land and water rights. Taxes were levied on the agricultural surplus, providing the revenue required to support the pharaoh's court, maintain the temples, and fund large-scale public works projects, such as the construction of pyramids and other monumental structures. Egyptian scribes played a critical role in recording these transactions, utilizing hieroglyphic writing and increasingly sophisticated mathematical techniques to calculate tax assessments and track revenue. The complex system of record-keeping reflects the importance of accurate accounting in maintaining the stability and prosperity of the Egyptian state.

The methods employed for assessing taxes varied across different ancient civilizations. In some cases, taxes were levied as a fixed percentage of agricultural output, while in others, they were based on the size of landholdings or the number of livestock owned. However, in all instances, accurate counting and record-keeping were essential for ensuring fairness and preventing abuses of power. The development of more precise counting systems, as well as the invention of more sophisticated mathematical tools, led to improvements in the efficiency and accuracy of taxation systems. These advancements were not merely technical improvements but had a significant impact on the political and social structures of these ancient societies.

The Roman Empire, known for its efficient administration and vast infrastructure, provides another compelling case study in the interplay between counting and governance. The Roman system of taxation was remarkably organized and far-reaching. Detailed censuses were conducted to accurately assess taxable properties and populations. This information was crucial for determining tax rates and ensuring equitable distribution of tax burdens. The Romans developed sophisticated accounting techniques to manage the vast revenue generated by their empire. Roman tax collectors were responsible for the collection of taxes across the empire, and their activities were carefully monitored to prevent corruption or mismanagement. The efficient collection of taxes provided the funds for the construction of roads, aqueducts, and other public works that were essential to the functioning of the Roman Empire. The system's success rested upon the ability of Roman officials to accurately count, record, and manage immense quantities of numerical data related to taxation.

Beyond the direct applications in taxation, counting systems played a crucial role in resource allocation within ancient cities. The distribution of food, water, and other essential resources required careful planning and effective management. The accurate recording of inventory levels allowed administrators to monitor supply levels and anticipate potential shortages. Efficient distribution systems, often utilizing networks of roads and canals, ensured that essential goods reached the population, contributing to social stability and minimizing the risk of famine or unrest. The precise tracking of resources, often facilitated by sophisticated numerical systems, was therefore an integral part of urban governance, underpinning the economic and social stability of ancient cities.

The development of more complex mathematical operations, such as multiplication and division, was closely intertwined with the demands of taxation and resource allocation. The need to calculate tax assessments on large quantities of goods or landholdings pushed mathematicians to devise more efficient methods of calculation. The invention of the abacus, for instance, facilitated more rapid and accurate calculations, significantly improving the efficiency of tax administration and resource management. The development of increasingly sophisticated writing systems, such as cuneiform and hieroglyphs, enabled the recording and dissemination of complex numerical data, facilitating communication and collaboration among administrators and tax collectors.

The relationship between taxation and resource allocation was not always straightforward. The level of taxation could affect economic productivity, and inefficient allocation of resources could lead to social unrest. Successful governance required a delicate balance between the need to generate revenue and the need to maintain economic stability and social harmony. The development of sophisticated counting systems and mathematical techniques played a vital role in achieving this delicate balance. Accurate accounting, efficient tax collection, and careful resource management were essential ingredients in the recipe for successful urban development and societal prosperity in the ancient world. The history of these ancient cities thus offers a powerful illustration of how the seemingly simple act of counting became a fundamental pillar of complex societal structures, demonstrating the profound and lasting impact of mathematics on the course of human civilization.

Trade and Commerce in Urban Centers

The burgeoning urban centers of antiquity witnessed not only the development of sophisticated governance structures but also the rise of complex and dynamic trading networks. The efficient functioning of these networks was inextricably linked to the development of increasingly refined counting systems. Trade, unlike the relatively localized transactions of smaller settlements, demanded a level of numerical precision and record-keeping far exceeding the needs of a simpler agrarian society. The ability to accurately account for goods exchanged, debts incurred, and profits accrued became essential for the economic vitality of these burgeoning cities.

One of the earliest examples of the application of counting to commerce can be found in the bustling markets of ancient Mesopotamia. The Sumerians, renowned for their entrepreneurial spirit and sophisticated economic practices, established extensive trade routes that connected their cities to distant regions. The exchange of goods, ranging from agricultural products and textiles to precious metals and exotic spices, required meticulous record-keeping. Clay tablets, the ubiquitous medium of Mesopotamian record-keeping, abound with evidence of commercial transactions. These tablets meticulously document quantities, prices, and the identities of buyers and sellers. The development of cuneiform script, with its ability to represent numerals and other symbolic information, was crucial in facilitating this complex system of commercial record-keeping. The consistent use of standardized weights and measures was also critical, ensuring fairness and transparency in transactions and minimizing the potential for disputes.

The standardization of weights and measures was not simply a matter of convenience; it was a fundamental element in the development of a stable and thriving market economy. Without standardized units, accurate pricing and fair exchange would have been impossible. The establishment of standardized units, meticulously calibrated and widely accepted, facilitated the smooth flow of goods and services, contributing to the economic prosperity of Mesopotamian cities. The meticulous nature of the records suggests that the Sumerians understood the importance of maintaining accurate and verifiable accounts, not just for individual transactions but also for tracking larger economic trends and managing the city's overall economic health. The emergence of specialized mercantile professions, including merchants, brokers, and accountants, reflects the growing complexity and importance of commercial activities within Mesopotamian society.

Ancient Egypt, similarly, possessed a vibrant trading economy facilitated by sophisticated counting systems. The Nile River, serving as a natural highway, connected various regions of the kingdom, allowing for the extensive exchange of goods and resources. Egyptian traders ventured far beyond the borders of their kingdom, establishing trade links with neighboring civilizations. The transportation and exchange of large quantities of goods required precise accounting to manage inventories, track shipments, and monitor profits. Hieroglyphic inscriptions, found on various artifacts and monuments, provide insights into the scale and complexity of Egyptian trade. These inscriptions often include numerical notations that represent quantities of goods, prices, and other relevant commercial information.

The development of increasingly sophisticated mathematical techniques in Egypt, particularly in areas like fractions and arithmetic progressions, suggests a direct correlation between the demands of commerce and advancements in mathematical thought. The ability to efficiently calculate the value of large quantities of goods, to determine fair exchange rates, and to manage complex financial transactions was essential for success in the competitive world of Egyptian trade. The establishment of official marketplaces and the regulation of commercial activities underscore the importance of commerce within Egyptian society. The state played a significant role in regulating trade, ensuring fair practices and preventing fraud, reflecting the crucial role of commerce in the Egyptian economy.

In the ancient Greek world, the growth of city-states, or poleis, fostered a flourishing commercial sector. The rise of democracy and the expansion of maritime trade created an environment where sophisticated financial instruments and practices emerged. The Greeks developed advanced systems of accounting and bookkeeping, utilizing various forms of writing to record financial transactions. The use of tally sticks, initially simple counting aids, evolved into sophisticated methods for recording credit and debt. These tally sticks, often notched to represent specific quantities or transactions, provided a reliable and verifiable record of financial dealings. The development of standardized coinage, featuring consistent weights and purity standards, facilitated more efficient and widespread trade. The introduction of standardized coinage, in contrast to bartering, significantly reduced the risks and complexities of transactions.

The development of banking systems in ancient Greece marked a pivotal moment in the history of commerce. Temples and wealthy individuals often served as depositories for valuables, effectively functioning as early banks. These institutions not only provided safe storage for money but also facilitated loans and credit transactions, laying the groundwork for more formal banking institutions. The use of standardized units of currency and the development of advanced accounting systems facilitated the expansion of credit and debt markets, contributing to the economic dynamism of Greek city-states. The emergence of sophisticated contractual agreements, detailing terms of payment and other legal aspects, reflects the increasing complexity of commercial interactions. These contracts, often inscribed on stone or clay tablets, played a critical role in ensuring fair and transparent transactions.

The Roman Empire, renowned for its vast infrastructure and extensive trade networks, offers another compelling example of the interplay between counting and commerce. The Roman system of taxation, already discussed, was intrinsically linked to the commercial activities within the empire. The vast quantities of goods traded throughout the empire necessitated sophisticated accounting practices. Roman merchants, often organized in guilds or associations, developed sophisticated methods for managing inventories, tracking shipments, and accounting for profits and losses. The development of advanced accounting systems, often using Roman numerals, enabled

Roman merchants to manage the complexities of large-scale trade. The vast network of Roman roads and sea lanes facilitated the efficient transportation of goods and helped expand the reach of Roman markets. The standardization of weights, measures, and currency contributed to the fairness and efficiency of transactions throughout the empire.

The establishment of markets and trading posts throughout the Roman Empire fostered economic growth and exchange. These markets served as hubs for the exchange of goods and services, often regulated by Roman authorities to ensure fair practices. The presence of large-scale infrastructure projects, such as aqueducts and roads, further facilitated commerce by providing vital services and infrastructure that supported trade. The Roman legal system played a vital role in regulating commercial activities, ensuring the enforcement of contracts and providing mechanisms for resolving disputes. The Romans, adept administrators, understood the crucial link between a well-regulated commercial sector and the overall prosperity of the empire.

In conclusion, the development of counting systems was not simply a matter of abstract mathematical progress but a fundamental prerequisite for the growth of urban centers and their complex commercial activities. From the clay tablets of Mesopotamia to the standardized coinage of Rome, the ability to accurately count, record, and manage numerical data underpins the economic dynamism of ancient cities. The evolution of sophisticated accounting systems, standardized units of measurement, and the emergence of banking systems all reflect the intertwining of mathematical advancements and economic prosperity. The study of these ancient systems offers invaluable insights into the relationship between counting, commerce, and the rise of complex societies. The sheer scale of trade and commerce in these ancient cities demanded precision, accuracy, and an ever-increasing sophistication in mathematical understanding; a testament to the enduring power of numbers to shape human civilization.

Social Organization and Governance

The intricate tapestry of urban life in ancient civilizations was not woven solely from economic threads; the very fabric of social organization and governance was deeply intertwined with the ability to count. While the previous sections highlighted the crucial role of counting in commerce, its influence extended far beyond the marketplace, shaping the structure of societies and the mechanisms of power. The capacity to enumerate, to quantify, and to record numerical data provided the tools necessary for maintaining social order, managing populations, and allocating resources in a way that profoundly impacted the development of complex urban environments.

One striking example of the profound impact of counting on governance can be found in ancient Rome. The Roman Empire, with its vast and diverse population, relied heavily on census data for administrative purposes. These censuses, conducted periodically throughout the empire's history, were not merely headcounts; they provided a comprehensive snapshot of Roman society, detailing not only the number of citizens but also their social status, property ownership, and tax obligations. This detailed numerical information was essential for the efficient functioning of the Roman state.

The Roman census served several vital functions. Firstly, it was fundamental to the system of taxation. Citizens were assessed taxes based on their wealth and property, as recorded in the census. The accuracy and impartiality of this process were paramount to the empire's financial stability. A reliable census ensured that tax burdens were distributed fairly, preventing widespread resentment and unrest. The Roman system of taxation was incredibly sophisticated for its time, showcasing a deep understanding of the relationship between numerical data and equitable resource allocation. The complexities of this system were reflected in the extensive bureaucratic apparatus necessary to collect and process the census data, underlining the significance of numerical literacy within the Roman administrative structure.

Moreover, the census played a crucial role in military recruitment. Rome's legions, the backbone of its military power, were drawn from the citizenry. The census provided the data needed to determine the number of eligible recruits and to ensure that conscription was carried out fairly and efficiently. This close linkage between census data and military organization highlights the vital role of numerical information in maintaining the security and stability of the empire. The detailed nature of the census allowed the Roman authorities to assess potential threats and allocate resources strategically to defend the vast frontiers of the empire. Without the capacity to accurately count and classify its population, Rome's military might would have been significantly hampered.

Beyond taxation and military recruitment, the census served other crucial administrative functions. It provided invaluable information for infrastructure planning, helping the Roman authorities to allocate resources efficiently and build the impressive network of roads, aqueducts, and public buildings that characterized the empire. The census also played a role in determining the distribution of public services, like grain supplies or public works programs, ensuring a level of social welfare that contributed to the empire's stability. The ability to anticipate societal needs and distribute resources strategically based on quantitative data was instrumental in maintaining social order and preventing widespread social unrest. This underscores the intricate connection between counting and the capacity of the Roman state to govern effectively a vast and diverse population.

Ancient Egypt offers another fascinating illustration of the link between counting and social organization. Egyptian society was highly stratified, with a rigid hierarchy ranging from the pharaoh at the apex to the peasant farmers at the base. The maintenance of this complex social order was inextricably linked to the ability to enumerate and categorize the population. While precise details about the methods of enumeration remain debated by scholars, hieroglyphic inscriptions and other artifacts reveal the extensive use of numerical notations to represent social categories and to track individuals within the hierarchical system.

The allocation of resources within ancient Egypt was heavily influenced by the numerical data available. The agricultural surplus, the very lifeblood of the Egyptian economy, was managed through complex systems of accounting that involved measuring land, tracking harvests, and ensuring the efficient distribution of grain and other essential resources. This complex system of resource management, again deeply rooted in numerical literacy, served to maintain social order and stability within Egyptian society. Disruptions to this meticulously balanced system could have easily led to social unrest and instability.

The construction of monumental structures, such as pyramids and temples, demanded an astonishing level of organizational skill and precision, both in terms of planning and execution. These ambitious projects required the mobilization of vast labor forces and the precise management of resources. Evidence suggests that Egyptian administrators made use of sophisticated counting systems to track the progress of these projects, allocating labor, materials, and time in a highly efficient manner. The very scale of these projects serves as testament to the sophisticated numerical capabilities of the ancient Egyptians and highlights the vital role of counting in their ability to manage large-scale projects and maintain a highly organized society.

The correlation between counting and the development of bureaucracies is undeniable. Bureaucracies, with their intricate systems of record-keeping and administration, are essentially reliant on the ability to handle and interpret large quantities of numerical data. The emergence of bureaucratic structures in ancient civilizations coincided with the development of increasingly sophisticated counting systems, suggesting a direct causal relationship between the two.

In Mesopotamia, the emergence of cuneiform script, with its ability to represent numbers, facilitated the development of complex administrative structures. Clay tablets abound with evidence of bureaucratic record-keeping, documenting taxation, land ownership, and other administrative details. The meticulous nature of these records demonstrates a high level of numerical literacy within Mesopotamian society and emphasizes the crucial role of counting in supporting the functioning of a centralized bureaucracy. This capacity for record-keeping was essential for maintaining social order and for managing the complex affairs of a city-state.

In ancient China, the development of sophisticated counting systems, including the use of counting rods and the invention of the abacus, were closely linked to the evolution of complex bureaucratic structures. The imperial bureaucracy played a central role in the administration of the vast Chinese empire, managing taxation, irrigation projects, and other essential aspects of governance. The ability to efficiently collect and analyze numerical data was critical to the effective functioning of this centralized administrative system. The detailed record-keeping systems employed in ancient China were central to maintaining the stability and longevity of the empire.

In conclusion, counting was not simply a tool for economic activity; it was the bedrock of social organization and governance in ancient cities. From the Roman census to the hierarchical structure of ancient Egypt, numerical data provided the essential framework for maintaining social order, regulating populations, and allocating resources. The development of sophisticated counting systems facilitated the rise of complex bureaucracies, capable of managing the intricate affairs of large and complex urban societies. The ability to count, record, and interpret numerical information was indispensable for the success and stability of these ancient civilizations, emphasizing the profound and far-reaching impact of counting on the development of human societies. The intricate interplay between counting, social organization, and governance reveals a fundamental truth: the ability to quantify the world around us has profoundly shaped the course of human history and civilization.

The Impact of Urbanization on Mathematical Development

The increasing complexity of urban life acted as a powerful catalyst for mathematical innovation. As cities grew, so too did the demands for efficient systems of administration, resource management, and construction. These practical necessities fueled the development of more sophisticated mathematical tools and techniques, creating a reciprocal relationship where the growth of cities spurred mathematical progress, and the advancements in mathematics, in turn, facilitated further urban expansion and complexity.

One of the most immediate impacts of urbanization on mathematics was the need for improved systems of numeration. Early agricultural villages may have managed with simple tallying systems, but the scale and diversity of urban economies necessitated more robust methods for representing and manipulating numbers. The development of positional numeral systems, such as the Babylonian sexagesimal system and the later Hindu-Arabic decimal system, were pivotal advancements that greatly simplified arithmetic operations. These systems, capable of representing arbitrarily large numbers with a limited number of symbols, allowed for efficient bookkeeping, taxation, and trade on an unprecedented scale. The sheer volume of transactions within a bustling city demanded efficient methods of calculation, driving innovation in numerical systems and computational techniques.

The administration of large urban populations also posed significant challenges, requiring effective methods for organizing and managing resources. The construction of monumental architecture, such as ziggurats, pyramids, and temples, was only possible through meticulous planning and the coordination of massive workforces. These ambitious projects required sophisticated geometrical knowledge, enabling the precise measurement of land, the calculation of volumes, and the accurate alignment of structures. The design and construction of irrigation systems, vital for supporting large urban populations, also necessitated a detailed understanding of geometry and surveying techniques. Evidence from ancient Egypt and Mesopotamia demonstrates a high level of geometrical understanding embedded within the design and execution of monumental architecture and intricate water management systems. The need to manage these complex projects spurred advancements in geometry and measurement, which subsequently found applications in other areas of urban life.

Furthermore, the growth of cities led to the development of more sophisticated accounting practices. The management of taxes, the distribution of goods, and the recording of financial transactions all demanded reliable and accurate record-keeping systems. The invention of writing, initially developed for keeping track of economic transactions, played a crucial role in this development. The ability to record and preserve numerical data allowed for better oversight of government operations, facilitated the growth of trade, and enabled the development of more complex financial instruments. The sophisticated accounting systems of ancient Mesopotamia, for instance, demonstrate the close link between urban development and the need for more advanced methods of numerical record-keeping. These advanced accounting practices facilitated efficient resource allocation, contributing to the overall stability and prosperity of the city.

The establishment of urban centers also played a significant role in the dissemination of mathematical knowledge. Cities acted as hubs for intellectual exchange, bringing together individuals from diverse backgrounds and fostering collaboration and innovation. Schools and academies, often associated with temples or religious centers, emerged as important sites for the transmission of mathematical knowledge. The concentration of scribes, mathematicians, and astronomers within urban centers created an environment conducive to the development and spread of mathematical ideas. The exchange of ideas and techniques across different regions and cultures contributed to the ongoing refinement and development of mathematical concepts and methods. The flow of mathematical knowledge, facilitated by trade routes and cultural exchange, created a dynamic environment where innovations could be shared, tested, and further developed. This dynamic interplay between different mathematical traditions fostered a more rapid pace of advancement than would have been possible in smaller, more isolated communities.

The impact of urbanization on mathematical development extended beyond the practical applications of mathematics. The increased complexity of urban life led to a growing demand for problem-solving skills, fostering the development of more abstract and sophisticated mathematical concepts. The challenges posed by urban administration, engineering, and commerce stimulated intellectual curiosity, pushing mathematicians to develop new theories and techniques. The abstract nature of mathematical inquiry was increasingly appreciated, leading to a greater focus on theoretical mathematics and its underlying principles. This shift towards more abstract thinking contributed to the overall advancement of mathematical knowledge and helped lay the foundations for future mathematical breakthroughs.

Specific examples further illuminate this reciprocal relationship. The development of sophisticated irrigation systems in Mesopotamia required detailed knowledge of geometry, surveying, and engineering. The intricate network of canals and ditches, essential for agricultural productivity, demanded precise calculations and careful planning. The design and construction of these systems, in turn, stimulated further advancements in geometry and mathematics. The ability to accurately measure land, calculate water flow, and design efficient water distribution systems was directly linked to the mathematical knowledge available to the engineers and administrators. The success of these projects, in turn, demonstrated the practical value of mathematical knowledge, fostering further investment in mathematical research and education.

In ancient Egypt, the construction of the pyramids, remarkable feats of engineering and planning, demanded sophisticated mathematical knowledge. The precise alignment of the pyramids, their massive scale, and the intricate internal chambers testify to the advanced geometrical and numerical skills of the Egyptian builders. The coordination of thousands of workers, the accurate measurement of materials, and the precise positioning of blocks all relied on advanced mathematical methods. The success of these ambitious projects, coupled with the enduring impact of these structures, highlighted the power of mathematical knowledge and the importance of mathematical skills in achieving ambitious goals. This in turn fueled further investment in the development and dissemination of mathematical skills within Egyptian society.

The Roman Empire, with its vast infrastructure and complex administrative systems, also provides a compelling example of the relationship between urbanization and mathematical development. The Roman road network, aqueducts, and public buildings were testaments to Roman engineering prowess, and their construction demanded a significant level of mathematical knowledge. Furthermore, the Roman census, a crucial tool for managing the vast empire, provided a wealth of demographic data that required sophisticated methods of data collection, analysis, and interpretation. The Roman system of taxation, based on property ownership, further required sophisticated numerical methods. The effectiveness of these administrative tools helped consolidate and maintain the vast Roman Empire.

In conclusion, the growth of cities acted as a powerful catalyst for mathematical development. The increasing complexity of urban life, with its demands for efficient resource management, accurate accounting, and ambitious construction projects, created a fertile environment for mathematical innovation. Urban centers served as hubs for the dissemination of knowledge, bringing together individuals from diverse backgrounds and fostering collaboration. The practical applications of mathematics in urban contexts demonstrated the value of mathematical knowledge, stimulating further investment in mathematical research and education. The reciprocal relationship between urbanization and mathematical development reveals a fundamental truth: the advancement of human societies and their ability to create complex urban environments have been inextricably linked to advancements in mathematical understanding and its practical application. This dynamic interaction has shaped the course of human history, propelling both mathematical and societal progress.

Chapter 5: Counting in the Medieval and Renaissance Periods

The Transmission of Mathematical Knowledge

The transmission of mathematical knowledge during the Medieval and Renaissance periods was a complex process, shaped by a confluence of factors including religious fervor, expanding trade networks, and the rediscovery of classical texts. Unlike the relatively centralized power structures of ancient empires, the dissemination of mathematical ideas during this era was characterized by a decentralized, multi-directional flow of information across diverse cultures and regions. The translation and adaptation of existing knowledge played a vital role, often leading to the refinement and expansion of established mathematical concepts.

The Islamic Golden Age (roughly 8th-13th centuries CE) played a pivotal role in preserving and expanding upon classical Greek mathematics. Scholars in Baghdad, Damascus, Cairo, and other centers of learning translated and studied the works of Euclid, Archimedes, Ptolemy, and other Greek mathematicians. This undertaking wasn't merely a passive act of preservation; it was an active engagement with these texts, leading to significant advancements in algebra, trigonometry, and number theory. Al-Khwarizmi, for example, is renowned for his work on algebra, introducing the concept of algorithms and systematic methods for solving equations. His treatise, *Kitab al-jabr wal-muqabala*, not only provided a clear exposition of algebraic techniques but also fundamentally shifted the focus from geometry-based problem-solving toward a more abstract and symbolic approach to mathematics. His influence on subsequent generations of mathematicians, both in the Islamic world and later in Europe, was profound and lasting. The development of Arabic numerals, a more efficient system than Roman numerals, also facilitated mathematical calculations and its eventual adoption throughout Europe had far-reaching consequences.

The role of translation houses and libraries, often associated with royal courts or religious institutions, was crucial to this process of knowledge transfer. These centers actively sought out and translated texts from various languages, including Greek, Sanskrit, and Persian. The translation process itself was not simply a literal rendering of one language into another; it involved interpretation, adaptation, and often the incorporation of existing mathematical traditions. This dynamic exchange resulted in the synthesis of different mathematical perspectives, enriching and expanding upon the existing body of knowledge. For example, the transmission of Indian mathematical concepts, particularly the decimal system and sophisticated methods of calculation, through Islamic channels to Europe profoundly impacted European mathematics.

The Crusades, despite their violent context, also inadvertently facilitated the transfer of mathematical knowledge. Contact between European Crusaders and the Islamic world exposed them to advanced mathematical techniques and ideas that were largely unknown in Europe. While the intellectual exchange was often indirect and uneven, it contributed to a gradual increase in mathematical sophistication within Europe. The return of Crusaders, carrying with them translated texts and knowledge of mathematical advancements, slowly contributed to shifting the European mathematical landscape. This flow, however, was certainly not one-sided. European scholars also contributed to the mathematical discourse, albeit more gradually and indirectly compared to the Islamic world's initial role.

The establishment of universities in Europe during the High Middle Ages marked a significant turning point in the transmission of mathematical knowledge. These institutions provided a structured environment for the study of mathematics, attracting students and scholars from across Europe. Monasteries also played an important role in preserving and transmitting knowledge, acting as repositories of ancient texts and centers of learning. Scribes diligently copied manuscripts, ensuring the survival of classical mathematical works. The intellectual climate within monasteries, often focused on theological and philosophical studies, inadvertently fostered an environment where mathematical exploration could flourish, particularly in areas connected to astronomy and calendrical calculations, essential for religious observances.

The Renaissance period witnessed a renewed interest in classical learning, fueled by the rediscovery of lost or neglected Greek and Roman texts. Scholars like Leonardo da Vinci, even though not strictly mathematicians, demonstrated a profound understanding of geometry and its applications in art and engineering. Their work highlights the interconnectedness of mathematical knowledge with other disciplines during this period. The invention of the printing press proved to be revolutionary in the dissemination of mathematical knowledge.

The ability to mass-produce books and other printed materials greatly increased the accessibility of mathematical texts, accelerating the spread of mathematical ideas across Europe. This made mathematical knowledge less reliant on the painstaking and labor-intensive process of manual copying and allowed for wider access and discussion, fostering a more rapid exchange of ideas.

The exploration and colonization of new lands during the Age of Exploration also inadvertently contributed to the dissemination of mathematical knowledge. Navigation, cartography, and astronomy played crucial roles in these voyages, requiring advanced mathematical skills. The demand for skilled mathematicians and navigators increased, resulting in the development of new mathematical tools and techniques. This, in turn, further accelerated the transmission of mathematical ideas across continents and fostered inter-cultural exchange. The process of mapping new lands and charting sea routes spurred advancements in trigonometry and geometry. The cross-cultural exchange resulted in the incorporation of indigenous mathematical practices into European systems of knowledge, leading to a richer understanding of mathematics itself.

Specific examples highlight these interconnected exchanges. Fibonacci's introduction of the Hindu-Arabic numeral system to Europe, through his influential book *Liber Abaci*, revolutionized European mathematics. The adoption of this system greatly simplified calculations, paving the way for significant advancements in algebra and arithmetic. This is a clear example of the power of cross-cultural transmission in driving significant mathematical progress. The transmission of mathematical knowledge was not confined to a linear progression; it was a multi-faceted process involving reciprocal exchange and adaptation.

Furthermore, the transmission was not simply confined to formal institutions. Trade networks, especially along the Silk Road, facilitated the exchange of mathematical ideas and techniques. Merchants, travelers, and scholars shared their knowledge as they traveled between different regions, contributing to the broader dissemination of mathematical ideas. This informal transmission of knowledge was arguably equally as crucial as the formal routes of universities and monasteries in shaping the mathematical landscape.

The transmission of mathematical knowledge during the Medieval and Renaissance periods was far from a uniform or unidirectional process. It was a dynamic interplay of cultural exchange, translation, adaptation, and innovation, shaped by political, religious, and economic factors. The rediscovery of classical texts, the rise of universities, the development of printing, and the expansion of trade networks all contributed to a rapid acceleration in the dissemination of mathematical ideas across diverse cultures. This transmission was not merely a transfer of information; it was a creative process of synthesis and adaptation, resulting in the enrichment and expansion of mathematical knowledge itself, laying the foundation for the development of modern mathematics. The story of mathematical transmission during these periods serves as a testament to the interconnectedness of human societies and the enduring power of knowledge to transcend geographical boundaries and cultural differences.

The Development of Arabic Numerals

The story of mathematics in the Medieval and Renaissance periods is inextricably linked to the evolution and spread of the Arabic numeral system. Before its widespread adoption in Europe, calculations were a cumbersome affair, reliant on the Roman numeral system. Roman numerals, with their reliance on additive and subtractive principles (e.g., IV for 4, IX for 9), lacked the positional notation that underpins the efficiency of the Arabic system. Multiplication and division, in particular, were significantly more challenging using Roman numerals, severely limiting the scope of complex mathematical operations. This inherent limitation hampered the development of advanced mathematical techniques and restricted the progress of scientific inquiry across various disciplines.

The Arabic numeral system, with its ten digits (0-9) and the crucial concept of positional value, offered a revolutionary alternative. Each digit's value depended not only on its inherent worth but also on its position within a number. This seemingly simple innovation dramatically simplified arithmetic operations. Consider the task of multiplying 456 by 789 using Roman numerals (CDLVI x DCCLXXXIX). The complexity becomes immediately apparent; even a moderately sized multiplication would require an inordinate amount of time and a very high degree of mental dexterity. In contrast, using Arabic numerals, the same calculation is straightforward, readily performed with relative ease even by individuals with limited mathematical training.

This efficiency wasn't merely a convenience; it was a catalyst for progress. The ability to perform complex calculations more rapidly and accurately opened doors to new mathematical explorations. Algebra, for example, greatly benefited from this system. The concise representation of numbers allowed for the development of more sophisticated algebraic manipulations and the solving of increasingly intricate equations. The positional system, with its zero acting as a placeholder, allowed mathematicians to express abstract concepts and represent quantities with far greater precision. This increased precision was vital in the growing fields of astronomy, navigation, and engineering.

The adoption of the Arabic numeral system wasn't a sudden event but a gradual process spanning centuries, influenced by a complex network of cultural exchange and transmission. While the precise origins of the system are still debated by scholars, it is widely believed to have developed in India and subsequently spread eastward towards the Middle East, where Islamic scholars played a crucial role in its dissemination and refinement. The works of Al-Khwarizmi, particularly his *Kitab al-hisab al-Hindi* (Book of Indian Calculation), were instrumental in introducing the system to the Islamic world. This text provided detailed explanations of the system's principles, along with algorithms for performing arithmetic calculations. The translation of this and other texts into Arabic established it as a significant text within the broader mathematical community.

The spread of Arabic numerals through the Islamic world was not simply a matter of transferring knowledge; it was a process of active engagement, adaptation, and innovation. Islamic mathematicians incorporated and refined the system, further improving its efficacy and making it even more suitable for wider application. This dynamic interaction led to the development of sophisticated mathematical techniques and tools that, in turn, spurred further advancements.

The transmission of the Arabic numeral system to Europe was similarly a complex process, involving multiple pathways and actors. The Crusades, while primarily military endeavors, facilitated indirect contact between European scholars and the Islamic world. This contact, albeit often fraught with conflict, exposed European scholars to the advanced mathematical techniques being utilized in the Middle East, including the Arabic numeral system. Commercial trade routes, particularly those connecting Mediterranean ports with the Levant and the Middle East, also played a crucial role. Merchants and travelers carried not only goods but also knowledge, contributing to the gradual dissemination of mathematical ideas and techniques.

A pivotal moment in the history of European mathematics occurred with the publication of Fibonacci's *Liber Abaci* (1202). This influential text introduced the Arabic numeral system to a wider European audience, demonstrating its advantages over the Roman system and providing detailed examples of its application. Fibonacci, having traveled extensively throughout the Mediterranean, recognized the system's superior efficiency and meticulously documented its principles and applications in his book. *Liber Abaci* served not only as a manual for merchants seeking to streamline their calculations but also as a catalyst for broader mathematical innovation within Europe. The book's widespread influence led to the gradual adoption of the Arabic system, though the transition was far from instantaneous. Resistance to change from those accustomed to the familiar Roman numerals certainly played a role.

The acceptance of the Arabic numeral system wasn't uniform across Europe. In some regions, its adoption was relatively swift, whereas in others, Roman numerals persisted for a considerable period. This uneven dissemination highlights the multifaceted nature of cultural change, where established practices and social inertia could significantly impede the rapid acceptance of even demonstrably superior systems. The printing press, invented in the mid-15th century, played a decisive role in accelerating the adoption process. The ability to mass-produce mathematical texts made the Arabic numeral system accessible to a far wider audience than ever before. This significantly reduced the reliance on laborious hand-copying of texts and created a more efficient system for disseminating mathematical knowledge.

The impact of the Arabic numeral system extended far beyond simple arithmetic. Its introduction significantly facilitated advancements in various scientific and technological fields. Astronomy, for instance, relied heavily on complex calculations involving large numbers. The ability to perform these calculations more efficiently using the Arabic system allowed astronomers to develop more accurate models of the celestial bodies and to advance their understanding of the cosmos. Similarly, navigation, which required precise calculations of latitude and longitude, benefited greatly from the system's efficiency.

The subsequent development of advanced mathematical concepts and techniques was intrinsically linked to the adoption of the Arabic numeral system. The emergence of new algebraic techniques, advanced calculus concepts, and the development of efficient algorithms for solving complex mathematical problems were all propelled by the efficiency and precision offered by positional notation and the use of zero as a placeholder. This facilitated the later development of sophisticated mathematical models used in fields like physics, engineering, and economics.

In conclusion, the development and adoption of the Arabic numeral system represents a pivotal moment in the history of mathematics. Its superiority over the Roman system was undeniable, revolutionizing calculations and facilitating the emergence of more sophisticated mathematical tools and techniques. The complex process of cultural exchange, spanning centuries and involving diverse actors and pathways, highlights the multifaceted nature of knowledge transmission and the enduring influence of mathematical innovations on scientific and technological progress. The Arabic numeral system was a cornerstone of the modern mathematical world, the foundation upon which much of our present-day mathematical understanding rests. The story of its development and adoption serves as a powerful reminder of the interconnectedness of human societies and the profound impact of seemingly simple innovations on the trajectory of human civilization.

Advances in Algebra and Geometry

The adoption of the Arabic numeral system, as detailed earlier, didn't merely streamline arithmetic; it acted as a powerful catalyst for advancements in higher mathematics, specifically algebra and geometry. The increased efficiency in computation freed mathematicians to grapple with more complex problems, pushing the boundaries of theoretical understanding and leading to practical applications across various fields.

Fibonacci's *Liber Abaci*, while primarily focused on the practical applications of the new numeral system, also contained problems and solutions that hinted at the emerging sophistication of algebra. His work wasn't limited to the simple arithmetic of merchants; he introduced problems that required more abstract reasoning and symbolic manipulation, features crucial to the development of algebraic thought. For instance, many of his problems involved finding unknown quantities within word problems, laying the groundwork for the later development of algebraic equations and their systematic solution. These problems, often couched in the language of commerce or practical scenarios, were deceptively complex, demanding a level of symbolic reasoning that went beyond mere number crunching. They served as a bridge between the practical concerns of the era and the emerging abstract power of algebra.

The development of symbolic algebra, however, was a gradual process. Early algebra relied heavily on rhetorical methods, where equations were written out in words rather than using symbols. This approach, while functional, was cumbersome and limited the complexity of problems that could be solved efficiently. The transition to symbolic algebra, where letters and symbols represented unknown quantities, significantly enhanced the ability to manipulate equations and solve more complex problems. This evolution wasn't a sudden leap but an incremental development, influenced by various mathematical traditions across the globe. The contributions of Islamic mathematicians, such as Al-Khwarizmi, were particularly significant. His *Kitab al-jabr wal-muqabala*, from which we derive the word "algebra," presented a systematic approach to solving equations, laying the groundwork for subsequent developments. Al-Khwarizmi's work addressed quadratic equations, a significant advance over simpler linear equations, and his methods were gradually refined and expanded upon by later scholars.

The shift from rhetorical to symbolic algebra represented a profound shift in mathematical thinking. The use of symbols allowed mathematicians to represent abstract concepts with greater precision and manipulate them in a more systematic way. This enhanced ability to manipulate symbols paved the way for the development of more sophisticated algebraic techniques, including the solution of cubic and quartic equations. The Italian Renaissance witnessed significant advancements in this area. Mathematicians such as Scipione del Ferro, Niccolò Tartaglia, and Girolamo Cardano made breakthroughs in solving cubic equations, which were previously considered unsolvable. Their solutions involved complex manipulations of algebraic expressions, showcasing the increasing power of symbolic algebra. Cardano, in his influential treatise *Ars Magna* (The Great Art), compiled and extended these results, contributing significantly to the development of algebraic techniques. The solutions involved intricate manipulations of radicals and the emergence of complex numbers, showing a growing level of mathematical sophistication and abstraction.

Geometry, too, underwent a significant transformation during this period. While the Euclidean geometry of classical antiquity remained a cornerstone, new approaches and innovations emerged. The rediscovery and translation of classical Greek texts, coupled with advancements in algebra, fueled progress in geometry. The integration of algebraic techniques into geometric problems allowed for a new level of analysis and solution. For example, the construction of geometric figures became increasingly intertwined with algebraic equations.

Problems that were previously tackled using solely geometric methods could now be formulated and solved using algebraic techniques, leading to a richer understanding and broader application. The work of mathematicians like François Viète exemplified this trend. Viète's introduction of symbolic algebra greatly improved the way geometric problems were approached, leading to significant advancements in the field.

The development of projective geometry, a significant departure from Euclidean geometry, marked another important milestone. Projective geometry deals with the properties of geometric figures that remain invariant under projection. This approach, pioneered by mathematicians such as Girard Desargues and Blaise Pascal, provided a new perspective on geometry and led to the discovery of new geometric relationships and theorems. Projective geometry is less concerned with the measurement of lengths and angles and more focused on the incidence relations between points, lines, and planes. This focus on incidence relations opened up new avenues for exploration and led to advancements in the understanding of geometric transformations and their properties.

The interaction between algebra and geometry, a key feature of this period, was not just a theoretical exercise. It had significant practical applications. Advances in surveying, cartography, and architecture were directly influenced by these developments. The precise measurement of land, the creation of accurate maps, and the design of intricate buildings all relied on the application of advanced geometric and algebraic principles. The construction of grand cathedrals, palaces, and other architectural marvels of the Renaissance era demonstrated the sophistication of geometric and algebraic knowledge. These ambitious projects demanded a deep understanding of geometry and often involved complex calculations requiring algebraic solutions to determine optimal designs, structural stability, and proportional aesthetics.

The application of mathematical concepts to navigation and astronomy further highlighted the importance of these advancements. The development of more accurate methods for determining latitude and longitude, vital for seafaring, demanded significant advances in both trigonometry (which combines algebra and geometry) and spherical geometry. The ability to accurately calculate positions at sea improved trade routes and expanded exploration, demonstrating the direct link between theoretical mathematical developments and their impact on practical human endeavors. Similarly, astronomy, which already had a strong mathematical basis, benefited immensely from the enhanced tools provided by advancements in algebra and geometry. The refinement of astronomical models, the ability to predict celestial events with greater accuracy, and the increasing sophistication of astronomical instruments all relied on these developments. This connection between mathematical theory and astronomical observation further reinforced the growing importance of mathematics in the intellectual landscape of the Medieval and Renaissance periods.

The interplay between theoretical and practical applications in this era is critical to understand. The development of algebra and geometry was not solely driven by abstract intellectual curiosity; it was also fueled by the practical demands of society. Conversely, the theoretical advancements in these fields had a profound impact on the practical applications in various professions and societal endeavors. This reciprocal relationship between theoretical advancement and practical application became a hallmark of the scientific revolution that followed, further emphasizing the central role of mathematics in shaping our understanding of the world.

In conclusion, the advancements in algebra and geometry during the Medieval and Renaissance periods marked a crucial transition in the history of mathematics. The adoption of the Arabic numeral system provided the foundation for increased computational efficiency, allowing mathematicians to tackle more complex problems and develop more sophisticated techniques. The integration of algebraic and geometric methods, the development of symbolic algebra, and the emergence of new geometric perspectives, such as projective geometry, transformed mathematical thinking and opened up new avenues for exploration. These developments, far from being isolated theoretical exercises, had a profound and far-reaching impact on various practical applications across various fields, underscoring the close connection between theoretical innovation and societal progress in this pivotal era. The legacy of this period continues to shape the mathematical landscape we inhabit today. The seeds of modern mathematics were sown during this era, demonstrating the cumulative and interconnected nature of scientific and mathematical progress.

The Impact of the Printing Press

The invention of the printing press in the mid-15th century stands as a watershed moment in the history of knowledge dissemination, and its impact on mathematics was profound and far-reaching. Before the advent of printing, the transmission of mathematical ideas relied heavily on laborious hand-copying of manuscripts. This process was slow, expensive, and prone to errors, limiting the accessibility of mathematical texts to a small, privileged elite – primarily scholars and wealthy patrons. The painstaking creation of each manuscript meant that mathematical knowledge spread slowly and unevenly across geographical regions and social strata. A single error in copying could perpetuate itself through multiple subsequent copies, potentially distorting or obscuring important mathematical concepts. The scarcity of manuscripts also meant that access to specific works was limited, hindering the development of new mathematical ideas. Scholars often had to travel great distances to consult original manuscripts, a journey often fraught with danger and expense, further restricting the free exchange of ideas.

The printing press dramatically altered this landscape. Its ability to reproduce texts quickly and accurately at a significantly lower cost democratized the access to mathematical knowledge. Suddenly, mathematical texts were no longer the exclusive domain of a small, privileged group. The printing press facilitated the mass production of books, making mathematical knowledge accessible to a wider audience, including students, teachers, and practitioners in various fields. This wider dissemination had a catalytic effect on mathematical progress, fostering collaboration, debate, and the rapid spread of innovations. The increased availability of mathematical texts spurred the development of mathematical education, allowing more individuals to acquire mathematical skills and participate in the advancement of the field.

One of the most immediate consequences of the printing press was the standardization of mathematical notation and terminology. Before the widespread adoption of printing, variations in notation and terminology were common, creating barriers to understanding and collaboration across different regions and schools of mathematical thought. Printed texts, however, helped establish consistent notations and terminologies, facilitating better communication and understanding among mathematicians. This standardization streamlined the process of learning mathematics and aided the development of new mathematical concepts. The uniformity achieved through printing was critical for fostering a more cohesive and unified mathematical community.

The printing press also played a vital role in the preservation of mathematical knowledge. Hand-copied manuscripts were vulnerable to damage, loss, and decay. The durability of printed books significantly reduced the risk of losing important mathematical texts. This ensured that mathematical knowledge accumulated over centuries was preserved and made readily accessible to future generations. The printed word secured a continuity in the development of mathematics that would have been impossible to maintain with the earlier methods of manuscript transmission.

Several specific examples highlight the pivotal role played by the printing press in disseminating mathematical knowledge. Regiomontanus's *De triangulis omnimodis* (On Triangles of Every Kind), a comprehensive treatise on trigonometry, published in 1533, exemplifies this impact. Before printing, such a detailed and influential work would have been accessible to only a handful of scholars. The printed edition made this seminal work widely available, influencing subsequent generations of mathematicians and significantly advancing the field of trigonometry, which was critical to advancements in astronomy, navigation, and surveying. Similarly, the publication of Luca Pacioli's *Summa de arithmetica, geometria, proportioni et proportionalità* (Everything about Arithmetic, Geometry, Proportions and Proportionality) in 1494, a compendium of mathematical knowledge of the time, proved immensely influential in spreading knowledge about arithmetic, algebra, and geometry. The sheer volume of its circulation, made possible by printing, facilitated the widespread adoption of new mathematical techniques and methods.

Another notable example is the dissemination of the work of Nicolaus Copernicus, whose *De revolutionibus orbium coelestium* (On the Revolutions of the Heavenly Spheres) published in 1543, proposed a heliocentric model of the solar system. While not strictly a mathematical text, the book's impact on astronomy hinged on sophisticated mathematical calculations and models. Printing ensured its widespread distribution, leading to significant debate and eventually revolutionizing our understanding of the cosmos. The mathematical arguments underpinning Copernicus's theory became accessible to a wider audience, stimulating further research and shaping the scientific revolution. The publication and widespread distribution of his work, a feat only made possible by the printing press, transformed the scientific landscape.

The impact of the printing press extended beyond the mere dissemination of complete books. The creation of pamphlets, broadsides, and smaller printed materials enabled the rapid exchange of mathematical problems, solutions, and news. Mathematical competitions and challenges became more frequent, encouraging the development of new techniques and the dissemination of innovative approaches. These smaller printed materials often showcased new mathematical puzzles and problems, stimulating engagement and fostering creativity. Mathematical societies and academies used printed materials to disseminate information and facilitate communication among members, promoting collaboration and accelerating the pace of mathematical discovery. The printing press helped create a dynamic and interconnected mathematical community that fostered innovation and progress.

Moreover, the availability of printed mathematical texts profoundly influenced mathematical education. Prior to the printing press, mathematical education relied heavily on oral instruction and the limited availability of hand-copied manuscripts. The printing press made it possible to produce textbooks and educational materials in large quantities, creating a more consistent and standardized approach to mathematics education. This standardization allowed for a more widespread and effective teaching of mathematical principles. The emergence of printed textbooks facilitated the spread of knowledge and made mathematical education more widely accessible. The consistent application of established mathematical practices through the use of printed materials fostered a more uniform understanding across various educational institutions.

In conclusion, the printing press's role in the development of mathematics during the Renaissance cannot be overstated. It was a transformative catalyst that fundamentally altered the dissemination, preservation, and teaching of mathematical knowledge. By making mathematical texts more accessible, affordable, and consistent, the printing press fostered a more vibrant and interconnected mathematical community, accelerating the pace of innovation and facilitating the scientific revolution. The impact extends far beyond the specific examples cited, illustrating the profound and lasting impact of this technological leap on the progress of human understanding and mathematical advancement itself. The printing press fundamentally changed the trajectory of mathematical history, democratizing access to knowledge and propelling the field forward at an unprecedented rate. Its influence is still felt today, shaping how we create, access, and share mathematical information.

Counting and the Scientific Revolution

The Scientific Revolution, spanning roughly from the 16th to the 18th centuries, was not merely a period of groundbreaking discoveries; it was a profound shift in the way humanity understood and interacted with the natural world. At the heart of this transformation lay a fundamental change in methodology: the embrace of quantitative methods and the increasing reliance on counting, measurement, and mathematical analysis. While counting had been a cornerstone of human civilization for millennia, its application within a systematic scientific framework marked a decisive turning point. The ability to quantify phenomena, to express observations in numerical terms, allowed scientists to move beyond qualitative descriptions and engage in rigorous, testable analysis. This transition was deeply intertwined with the development of new mathematical techniques, creating a powerful synergy between mathematics and the various scientific disciplines.

One of the most significant areas where counting played a transformative role was astronomy. The ancient Greeks had made substantial contributions to astronomy, but their models were often based on geometric considerations rather than precise numerical calculations. The Scientific Revolution saw a dramatic shift towards a more quantitative approach. Nicolaus Copernicus's heliocentric model, while revolutionary in its conceptual shift, relied heavily on meticulous astronomical observations and complex mathematical calculations to predict planetary positions. While Copernicus's work was groundbreaking, it was Johannes Kepler who truly harnessed the power of quantitative analysis to refine and solidify the heliocentric model. Kepler's three laws of planetary motion, derived from painstaking analysis of Tycho Brahe's extensive observational data, demonstrated the predictive power of mathematical formulations in astronomy. The meticulous counting and recording of planetary positions, meticulously documented over years, formed the empirical bedrock upon which Kepler built his groundbreaking laws. His work exemplified the increasingly crucial role of quantitative data in achieving scientific breakthroughs. Kepler's reliance on numerical precision, meticulously tracking and analyzing observational data, marked a stark departure from previous qualitative astronomical models, and helped solidify the heliocentric model.

The contributions of Galileo Galilei further highlighted the importance of counting in the Scientific Revolution. Galileo's experiments with falling bodies, while seemingly simple, represented a crucial move towards a more quantitative understanding of physics. By carefully measuring the time it took for objects of different weights to fall, he challenged Aristotelian physics and laid the groundwork for Newton's laws of motion. His precise measurements, often using ingenious methods of time-keeping, provided empirical evidence to support his theories, pushing the field of physics toward a more mathematically rigorous framework. The careful recording and analysis of his experimental data, expressed in numerical terms, constituted a powerful argument against long-held assumptions about the physical world. Galileo's experiments were not mere observations; they were carefully designed experiments where numerical data played a crucial role in the validation of his hypotheses.

The development of calculus by Isaac Newton and Gottfried Wilhelm Leibniz marked another critical juncture. Calculus provided a powerful mathematical tool for analyzing continuous change and motion, which were essential for understanding phenomena in physics and astronomy. The ability to calculate instantaneous rates of change, through the process of differentiation, and to determine areas and volumes, via integration, revolutionized the ability to model and understand physical processes. While the conceptual foundations of calculus were rooted in geometrical insights, its practical application rested on the ability to perform calculations, often involving intricate numerical estimations and approximations. The development and application of calculus depended fundamentally upon the ever-increasing capacity to perform increasingly complex calculations.

The interplay between mathematics and other sciences during the Scientific Revolution was not a one-way street. The demands of scientific inquiry often spurred advancements in mathematics itself. For instance, the need to analyze astronomical data led to refinements in trigonometry and the development of new computational techniques. The quest for accurate navigation also drove innovations in cartography and geometry, demonstrating a powerful feedback loop between scientific needs and mathematical development. This mutual influence between science and mathematics was crucial in driving the scientific revolution forward.

Beyond astronomy and physics, counting and quantitative analysis permeated other scientific fields. In biology, the use of quantitative methods became increasingly important, particularly with the rise of observational studies and the development of taxonomy. The detailed counting of specimens, the careful measurement of anatomical features, and the statistical analysis of biological data became crucial for understanding biological diversity and evolutionary relationships. Likewise, in chemistry, the development of stoichiometry—the quantitative study of the proportions of reactants and products in chemical reactions—required precise measurements and calculations, solidifying chemistry's transformation from an alchemical art to a precise science.

The rise of probability and statistics further exemplifies the increasing importance of counting during the Scientific Revolution. The development of probability theory, driven partly by the need to analyze gambling games, provided new tools for analyzing uncertain events and making predictions based on limited data. This methodology found applications in various fields, from astronomy to medicine, enabling scientists to draw meaningful conclusions from noisy or incomplete data. The ability to quantify uncertainty and manage risk became increasingly crucial as scientific methodologies evolved.

The Scientific Revolution was characterized not only by major discoveries but also by a new way of thinking—a shift toward a more empirical and quantitative approach. Counting, measurement, and mathematical analysis emerged as essential tools for understanding the natural world. This wasn't a sudden revolution, but a gradual process where counting, initially a basic human activity, transformed into a cornerstone of the scientific method, allowing for rigorous testing and the formulation of precise, predictive models of the universe. The legacy of this period continues to shape our scientific endeavors today, underscoring the enduring importance of quantitative methods in scientific discovery and technological advancement. The ability to quantify, to translate observations into numerical data and to analyze that data using mathematical tools, became fundamental to the scientific method, continuing to influence and shape our understanding of the world. This powerful synergy between counting and scientific inquiry ultimately redefined the relationship between humanity and the cosmos.

Chapter 6: The Age of Reason and the Enlightenment

The Rise of Calculus and its Applications

The Scientific Revolution's emphasis on quantitative analysis found its most powerful expression in the development of calculus. This revolutionary mathematical tool, independently conceived by Isaac Newton and Gottfried Wilhelm Leibniz in the late 17th century, provided a framework for understanding continuous change and motion, phenomena previously resistant to precise mathematical description. Before calculus, mathematicians grappled with problems involving curves and instantaneous rates of change using cumbersome geometric methods. Newton and Leibniz, building upon the work of earlier mathematicians like Fermat and Barrow, introduced the concepts of differentiation and integration, providing elegant and powerful tools to tackle these challenges. Differentiation allowed for the determination of the instantaneous rate of change of a function, crucial for understanding velocity, acceleration, and other dynamic processes. Integration, conversely, provided a method for calculating areas under curves, volumes of solids, and the accumulation of quantities over time, essential for applications in physics, engineering, and beyond.

Newton's development of calculus was intimately linked to his groundbreaking work in physics. His *Principia Mathematica*, published in 1687, laid the foundation for classical mechanics, and calculus was the essential mathematical language through which his laws of motion and universal gravitation were expressed. Consider, for example, the trajectory of a projectile. Before calculus, predicting its path required complex geometric approximations. Newton, using his newly developed calculus, could precisely describe the projectile's motion at any given instant, factoring in gravity, initial velocity, and air resistance (though the latter required further mathematical advancements). This capacity to precisely model motion revolutionized physics and engineering, allowing for more accurate predictions and designs in everything from cannon trajectories to bridge construction.

The law of universal gravitation itself relies heavily on calculus. Newton's formulation, stating that every particle attracts every other particle in the universe with a force proportional to the product of their masses and inversely proportional to the square of the distance between their centers, is not merely a statement; it's a differential equation. Solving this equation, using the tools of calculus, allows us to predict the orbits of planets, the tides, and the movements of celestial bodies with remarkable accuracy. Before calculus, understanding these motions was a realm of approximation and speculation. Newton's calculus transformed it into a realm of precise prediction.

Leibniz, while developing calculus independently, approached the subject from a somewhat different perspective, emphasizing the concept of infinitesimals – infinitely small quantities. While the philosophical underpinnings of infinitesimals caused some debate among mathematicians at the time (and continue to be a subject of rigorous mathematical scrutiny today), Leibniz's notation proved significantly more adaptable and user-friendly than Newton's, contributing to the widespread adoption of calculus across Europe. His notation, still used today, makes the processes of differentiation and integration more transparent and accessible to students and researchers alike. This accessibility played a significant role in the dissemination and subsequent application of calculus across multiple scientific fields.

The applications of calculus extended far beyond celestial mechanics. In optics, it facilitated the precise calculation of lens shapes and the paths of light rays, leading to improvements in telescope design and other optical instruments. In fluid dynamics, calculus proved essential for modeling the flow of liquids and gases, crucial for understanding everything from blood circulation in the human body to the design of efficient water wheels and turbines. The study of heat transfer and thermodynamics, vital fields of inquiry during the Enlightenment, similarly benefited greatly from the application of calculus. The concepts of heat flow, temperature gradients, and thermal equilibrium all found precise mathematical expressions through calculus.

The development of calculus also profoundly impacted the field of engineering. The design and construction of bridges, buildings, and other structures demanded increasingly precise calculations to ensure stability and safety. Calculus provided the mathematical framework for understanding stress, strain, and structural dynamics, allowing engineers to design stronger and more reliable structures. The ability to calculate the bending moments, shearing forces, and deflections within a structure became crucial for optimizing designs and ensuring safety, a shift that contributed significantly to the architectural marvels and robust infrastructure developed throughout the Enlightenment and beyond.

Beyond its immediate applications, calculus's influence extends to the broader philosophy of science. Its ability to model continuous change underscored the dynamic nature of the universe, moving away from the static worldviews of earlier eras. The precise mathematical description of motion and change fostered a more quantitative and predictive understanding of the natural world, solidifying the empirical approach to scientific inquiry that defined the Enlightenment. The ability to not just observe but to precisely measure and mathematically predict the behavior of systems became a cornerstone of the scientific method, profoundly altering how humanity approached the investigation of the universe.

Furthermore, the development of calculus spurred further mathematical advances. The need to solve increasingly complex differential equations led to the development of new mathematical techniques and concepts, creating a feedback loop between theoretical mathematics and its practical application in science and engineering. The development of numerical methods for approximating solutions to differential equations, for example, became crucial in situations where analytical solutions proved intractable. These numerical approaches, often involving iterative calculations, underscored the continued reliance on counting and computation, demonstrating the continuous interplay between the theoretical and practical aspects of mathematics.

The rise of calculus was not merely a culmination of prior mathematical knowledge; it was a catalyst for the future development of mathematics and science. Its impact extended far beyond the specific problems it solved, influencing the very way scientists approached the investigation of natural phenomena. The ability to represent continuous change mathematically transformed the study of motion, physics, engineering, and countless other fields, marking a fundamental shift in scientific understanding that persists to the present day. The development of calculus was, in essence, a pivotal moment in the evolution of counting, demonstrating the profound power of mathematical tools in unlocking the secrets of the universe and shaping the course of human civilization. The continuous evolution of calculus and its applications continues to be evident in the enduring power of this mathematical invention. The story of calculus is far from over; it's a story of continuous development, refinement, and ever-expanding applications, a demonstration of its foundational role in modern science and technology. Even today, new mathematical techniques and computational power continue to expand the range and depth of its applications, ensuring its continuing relevance in the pursuit of scientific knowledge.

Probability and Statistics: Quantifying Uncertainty

The triumph of calculus in providing precise mathematical descriptions of continuous change during the Enlightenment naturally led to a parallel quest: quantifying uncertainty. While calculus dealt with deterministic systems, where cause and effect were clearly linked, many phenomena in the real world were inherently probabilistic. The outcomes of events were not predictable with certainty, but rather occurred with varying likelihoods. This realization spurred the development of probability and statistics, fields that would prove indispensable for the advancement of scientific inquiry and numerous practical applications.

The roots of probability theory can be traced back to games of chance. The analysis of dice rolls and card games, seemingly frivolous pursuits, laid the foundation for a rigorous mathematical framework for understanding randomness. Early mathematicians, grappling with the seemingly unpredictable nature of these games, began to recognize patterns and regularities. Gerolamo Cardano, a 16th-century Italian polymath known for his work in mathematics and medicine, made significant contributions to the early understanding of probability, though his work remained largely unpublished during his lifetime. He explored the probabilities associated with different outcomes in games of chance, laying some of the groundwork for future developments.

However, it was in the 17th century that probability theory began to emerge as a distinct branch of mathematics. Pierre de Fermat and Blaise Pascal, two titans of mathematics, engaged in a correspondence that is now considered a pivotal moment in the history of probability. Their collaboration, spurred by a gambling problem posed by a French nobleman, led them to develop fundamental concepts, such as the calculation of expected values and the determination of probabilities for various events. Their insights laid the foundation for the formalization of probability theory as a mathematical discipline.

The next major step was the publication of *Ars Conjectandi* (The Art of Conjecture) by Jacob Bernoulli in 1713, posthumously published by his nephew. This monumental work offered a systematic treatment of probability theory, exploring concepts like the law of large numbers. Bernoulli's law of large numbers states that as the number of trials of a random event increases, the observed frequency of an outcome converges towards its theoretical probability. This provided a crucial link between theoretical probability and empirical observation, bolstering the credibility and applicability of probabilistic reasoning.

The 18th century witnessed further advancements in probability theory, notably through the work of Abraham de Moivre. De Moivre's contributions, particularly his development of the normal distribution, proved to be of paramount importance. The normal distribution, often depicted as a bell curve, describes the distribution of many natural phenomena. Its ubiquity in diverse fields, from measuring human height to analyzing experimental errors, speaks to the power and generality of this statistical concept. De Moivre's work paved the way for the application of probability to a wider range of scientific problems.

Thomas Bayes, an 18th-century Presbyterian minister, made a significant, albeit initially less recognized, contribution with his work on Bayesian inference. Bayes' theorem provides a framework for updating probabilities based on new evidence. While initially overlooked, this theorem has become a cornerstone of modern statistical inference, finding wide-ranging applications in fields such as machine learning, artificial intelligence, and medical diagnosis. The theorem allows for the incorporation of prior knowledge or beliefs into the probabilistic analysis, leading to more nuanced and informed inferences.

The development of probability was closely intertwined with the rise of statistics. Statistics, in its broadest sense, deals with the collection, analysis, interpretation, presentation, and organization of data. While the origins of statistical methods can be traced back to ancient civilizations' need for censuses and tax collection, the Enlightenment saw a formalization and refinement of these methods. The increasing availability of data, coupled with the development of mathematical tools for analyzing them, propelled the growth of statistics as a distinct discipline.

The application of probability and statistics extended far beyond games of chance. In astronomy, for instance, these tools allowed scientists to analyze observational data, account for measurement errors, and make inferences about the nature of celestial bodies. The precise determination of planetary orbits, the study of stellar distributions, and the analysis of astronomical observations all benefited immensely from statistical methods. Similarly, in the emerging field of demography, statistics played a crucial role in understanding population trends, mortality rates, and other demographic variables.

In the biological sciences, the application of statistics was transformative. The study of biological variation, the analysis of experimental results, and the testing of hypotheses all relied on statistical techniques. The rise of experimental biology, fueled by the Enlightenment's emphasis on empirical observation, created a significant demand for robust statistical methods to analyze data and draw meaningful conclusions. Statistical analysis allowed researchers to differentiate between random variation and true biological effects, paving the way for more precise and reliable scientific findings.

The philosophical implications of quantifying uncertainty were also profound. The ability to assign probabilities to events, while acknowledging the inherent limitations of deterministic knowledge, represented a significant shift in how scientists viewed the world. The acceptance of uncertainty did not signal a retreat from reason; instead, it broadened the scope of scientific inquiry to include phenomena that were inherently unpredictable.

The development of probability and statistics wasn't a linear progression. There were debates, revisions, and controversies surrounding fundamental concepts. Philosophical questions regarding the interpretation of probability, frequentist versus Bayesian interpretations, remain a topic of discussion even today. Nevertheless, the Enlightenment witnessed the emergence of a powerful mathematical framework for dealing with uncertainty, one that profoundly impacted the development of science, technology, and the broader human understanding of the world. The influence of these fields extends far beyond the scientific realm; their principles underpin decision-making processes in various fields, from finance and insurance to healthcare and public policy. The ability to quantify and manage risk became an indispensable aspect of modern life, a direct consequence of the intellectual ferment of the Enlightenment.

The continued development of probability and statistics during the Enlightenment and beyond is evident in their enduring importance in scientific understanding. New statistical techniques continue to be developed to address increasingly complex problems, and computational advances have made it possible to analyze vast quantities of data with unprecedented speed and accuracy. The legacy of the Enlightenment's contributions to probability and statistics continues to shape our world, as we rely more and more on data-driven decisions in all aspects of our lives. The ability to analyze data effectively, to quantify uncertainty, and to make informed judgments based on probability—these are all legacies of the Age of Reason and their impact is still unfolding. This capacity for understanding uncertainty, a direct outgrowth of the mathematical breakthroughs of the period, is a cornerstone of the modern scientific method and underpins countless advancements in our understanding of the world. The story of probability and statistics is far from concluded; it remains a vibrant and evolving field, continuously refining its tools and extending its reach into ever more complex domains of human knowledge.

The Development of Number Theory

The growing interest in probability and statistics during the Enlightenment naturally fostered a renewed focus on number theory, a field seemingly less concerned with the unpredictable nature of the world but nonetheless crucial for underpinning many mathematical structures. While seemingly abstract, the study of the properties of numbers and their relationships proved unexpectedly fertile ground for groundbreaking discoveries, providing elegant solutions to problems that had vexed mathematicians for centuries and laying the groundwork for critical applications in areas as diverse as cryptography and computer science.

One of the central figures in this resurgence of interest in number theory was Pierre de Fermat, whose contributions extended far beyond his correspondence with Pascal on probability. While famously known for his "Last Theorem," which remained unproven for over three centuries, Fermat's work in number theory was prolific and profoundly influential. His contributions were often presented as concise statements, often without detailed proofs, leaving a trail of challenging problems for future mathematicians to grapple with. This practice, while frustrating to some, significantly stimulated mathematical progress by encouraging others to explore his ideas, refine his methods, and ultimately expand the field.

Fermat's work on modular arithmetic, a branch of number theory concerned with the remainders of numbers after division by a fixed integer, laid the foundation for many subsequent advancements. His exploration of congruences, specifically the properties of numbers congruent modulo a particular integer, provided a powerful tool for analyzing number theoretic properties. This concept has far-reaching consequences, forming the backbone of numerous algorithms and cryptographic techniques used today. For instance, understanding Fermat's Little Theorem, which states that if p is a prime number, then for any integer a, the number $a^p - a$ is an integer multiple of p, is essential for modern cryptography. This theorem underpins primality tests, used to determine whether a given large number is prime – a crucial aspect of securing digital communications.

Another monumental figure in the development of number theory during the Enlightenment was Leonhard Euler, a Swiss mathematician whose prodigious output spanned numerous areas of mathematics and physics. Euler's contributions to number theory were particularly significant, building upon and extending the foundations laid by Fermat. Euler's work delved deeper into the properties of prime numbers, exploring their distribution and their relationships with other numbers. He made significant progress in understanding the distribution of primes, a notoriously difficult problem that continues to fascinate mathematicians today. His insights into the structure of prime numbers provided crucial tools for tackling a variety of number theoretic problems.

Euler's exploration of Diophantine equations, algebraic equations where only integer solutions are sought, significantly advanced the field. Diophantine equations, named after the ancient Greek mathematician Diophantus of Alexandria, pose a challenge: find integer solutions that satisfy given algebraic equations. Euler demonstrated remarkable ingenuity in tackling these problems, finding elegant solutions and developing techniques that continue to be used today. His work on the equation $x^2 + y^2 = z^2$, which relates to Pythagorean triples (sets of integers satisfying the Pythagorean theorem), is a prime example of his elegant and impactful contributions. His systematic exploration of these equations, revealing their underlying structures and providing methods for finding solutions, showcased the elegance and power of number theoretic techniques.

Euler's work also significantly impacted the development of analytic number theory, a branch that utilizes techniques from analysis (calculus and related fields) to tackle problems in number theory. This innovative approach combined the rigor of number theory with the power of continuous mathematics, leading to deeper insights into the distribution of prime numbers and other number theoretic concepts. His contributions here were particularly significant in pushing beyond the purely algebraic approaches that had characterized much of earlier number theory, opening up a rich landscape of new questions and methods for exploration.

The work of Fermat and Euler, though separated by time and geographic location, profoundly impacted the trajectory of number theory. Their discoveries and the problems they posed spurred a generation of mathematicians to probe deeper into the intricacies of numbers. They showcased the surprising connections between seemingly abstract mathematical concepts and their practical applications. Their legacies continue to resonate in contemporary mathematics, as the foundations they built continue to underpin modern research in number theory, cryptography, and computer science.

The impact of these Enlightenment-era advancements extended beyond the purely academic. The development of number theory significantly impacted the emerging field of cryptography. Before the Enlightenment, cryptographic techniques were relatively rudimentary, often relying on simple substitution ciphers or transpositions that were relatively easy to break. The advancement of number theory, particularly the deeper understanding of prime numbers and modular arithmetic, provided the mathematical foundations for creating vastly more secure cryptographic systems. The development of public-key cryptography in the 20th century, for example, relies heavily on number theoretic concepts like the difficulty of factoring large numbers into their prime components. The secure transmission of information across digital networks, essential for e-commerce, online banking, and many other aspects of modern life, rests on these mathematical foundations.

Furthermore, the study of number theory contributed to the development of algorithms and computational techniques. Number theoretic problems often require efficient methods for solving complex equations or finding patterns in sequences of numbers. This led to the development of sophisticated algorithms, many of which have found application in computer science and other areas. For example, algorithms based on number theoretic principles are used in computer graphics, digital signal processing, and other applications where efficient manipulation of large numbers is crucial.

The Enlightenment's focus on reason and empirical observation profoundly impacted the development of number theory. The emphasis on rigorous proofs and the exploration of abstract mathematical structures were hallmarks of this intellectual revolution. Mathematicians of the Enlightenment didn't just seek to solve problems; they sought to understand the underlying structure and logic of numbers, establishing the foundational principles that continue to shape the field today.

However, the story of number theory during the Enlightenment wasn't without its challenges. Many problems posed by Fermat and Euler remained unsolved for centuries, driving further research and innovation. The search for solutions to these seemingly simple problems pushed the boundaries of mathematical knowledge, leading to the development of new techniques and perspectives. This dynamic interaction between problems, solutions, and the development of new mathematical tools characterizes the progress of the field. The challenges posed by number theory continue to inspire mathematicians to this day, demonstrating the enduring legacy of the Enlightenment's contributions to the field. The legacy of Fermat and Euler, and their contributions to number theory, continues to resonate in the modern mathematical landscape, illustrating the lasting impact of the Age of Reason on our understanding of the fundamental properties of numbers and their applications in a vast array of fields. The development of number theory during the Enlightenment was not merely an abstract exercise; it laid the groundwork for many crucial advancements in technology and security, showcasing the power of pure mathematics to shape our world in unexpected and profound ways. The exploration continues, as mathematicians strive to unravel the deeper mysteries of numbers, driven by the same spirit of inquiry and intellectual curiosity that characterized the Age of Reason.

The Influence of the Enlightenment on Mathematical Thought

The Enlightenment, a period characterized by a fervent embrace of reason, empiricism, and individual liberty, profoundly reshaped the intellectual landscape of Europe and, consequently, profoundly influenced the development of mathematical thought. The emphasis on logical deduction, systematic inquiry, and the pursuit of universal truths resonated deeply within the mathematical community, fostering a shift towards a more rigorous, formalized, and axiomatic approach to the field. Before the Enlightenment, mathematical practices often relied on intuition, geometrical insights, and a less systematic approach to proof. While significant advancements had been made, the foundations of mathematical reasoning were often less explicitly articulated. The Enlightenment, however, ushered in an era of heightened scrutiny and a demand for greater precision in mathematical arguments.

This paradigm shift is vividly illustrated by the growing interest in the axiomatic method. Euclid's *Elements*, a cornerstone of classical geometry, had long served as a model of logical structure, but the Enlightenment saw a renewed appreciation for its rigorous approach. Mathematicians began to strive for similar levels of rigor in other branches of mathematics, seeking to establish clear axioms, fundamental, self-evident truths, upon which entire systems of theorems could be logically constructed. This pursuit of axiomatic foundations extended beyond geometry to encompass algebra, calculus, and other developing fields, laying the groundwork for the more abstract and formally defined mathematics of the 19th and 20th centuries.

The influence of Enlightenment ideals extended beyond the formal structure of mathematics, shaping the very philosophy of the subject. The emphasis on reason and logic challenged existing authorities and encouraged a more critical approach to mathematical knowledge. Mathematicians began to question accepted truths, demanding rigorous justification for even the most fundamental principles. This critical spirit spurred new lines of inquiry and led to a more nuanced understanding of the foundations of mathematics. The pursuit of certainty, a hallmark of Enlightenment thinking, became a driving force in mathematical research, leading to an increased focus on establishing the validity of mathematical statements beyond mere intuition or empirical observation. The quest for rigorous proof became a central feature of the mathematical enterprise.

One notable example of the Enlightenment's influence on mathematical practice is the development of calculus. While Newton and Leibniz independently developed the fundamental principles of calculus in the late 17th century, its foundations remained somewhat shaky. The Enlightenment witnessed a concerted effort to rigorously establish the theoretical basis of calculus, addressing concerns about infinitesimals and the legitimacy of certain mathematical operations. Mathematicians like Augustin-Louis Cauchy, in the early 19th century, played a pivotal role in this endeavor, developing a more rigorous approach based on the concept of limits, providing a firmer foundation for the seemingly paradoxical concepts of calculus.

The Enlightenment also fostered a greater emphasis on the relationship between mathematics and the natural world. The prevailing scientific spirit of the era, driven by observation and experimentation, influenced mathematical inquiry. Mathematicians began to seek mathematical models to describe natural phenomena, leading to the development of mathematical physics and a deeper integration between mathematics and the sciences. The successes of Newtonian mechanics, which relied heavily on mathematical tools like calculus and differential equations, further solidified this link, reinforcing the value of mathematical modeling in understanding the physical world. This interaction was not unidirectional; the challenges posed by physical problems often spurred the development of new mathematical concepts and techniques.

The Enlightenment's emphasis on precision and systematic thinking also impacted the organization and dissemination of mathematical knowledge. The creation of academic societies and journals provided a platform for mathematicians to share their work, engage in collaborative research, and subject their findings to peer review. This facilitated the rapid dissemination of new ideas and fostered a more collaborative environment for mathematical research. The establishment of institutions dedicated to the advancement of mathematical knowledge played a critical role in furthering the development of the field.

Beyond the specific advancements in various mathematical disciplines, the Enlightenment fostered a change in the social perception of mathematics. Mathematics transitioned from being viewed primarily as a tool for practical applications, such as accounting and surveying, to a subject worthy of study in its own right, for its inherent intellectual value. This elevation of mathematics to a discipline of pure intellectual pursuit encouraged greater exploration of abstract concepts and theoretical structures. The Enlightenment's emphasis on the inherent worth of knowledge for its own sake broadened the appeal of mathematics and fostered a more profound appreciation for its beauty and elegance.

However, the Enlightenment's impact on mathematics was not without its limitations. The quest for certainty and rigor led to some overly cautious approaches, potentially hindering the development of certain areas of mathematics. The emphasis on axiomatic systems, while beneficial in many respects, also led to prolonged debates about the foundations of mathematics in the later 19th and early 20th centuries. The search for self-evident truths proved more elusive than initially hoped, leading to unforeseen complexities and challenges in defining the very basis of mathematical knowledge.

Despite these complexities, the overall impact of the Enlightenment on mathematical thought was undeniably transformative. The period witnessed a profound shift in mathematical philosophy and practice, characterized by a greater emphasis on rigor, precision, the axiomatic method, and the interconnectedness between mathematics and the natural world. The Enlightenment's legacy continues to shape the development of mathematics today, influencing the way we approach mathematical problems, construct arguments, and evaluate the validity of mathematical statements. The spirit of critical inquiry, systematic reasoning, and the pursuit of universal truths, central tenets of the Enlightenment, remain cornerstones of the mathematical enterprise, shaping the ongoing exploration of the field's intricate and ever-evolving landscape. The rigorous standards and analytical frameworks established during this era laid the groundwork for the subsequent explosion of mathematical knowledge in the centuries that followed, continuing to inspire and guide mathematicians to this day. The Enlightenment's embrace of reason and its demand for clarity fundamentally altered the course of mathematics, shaping it into the sophisticated and powerful discipline we know today.

Counting and the Expansion of Scientific Knowledge

The Enlightenment's emphasis on reason and empirical observation profoundly impacted scientific inquiry, intertwining it inextricably with the power of counting. No longer were scientific observations merely qualitative descriptions; they became increasingly quantitative, relying on precise measurements, systematic data collection, and sophisticated mathematical analysis. This shift marked a fundamental transformation in the way scientists approached the natural world, propelling scientific progress across a broad spectrum of disciplines.

Astronomy, for instance, underwent a dramatic revolution during this period. The meticulous observations of celestial bodies, meticulously recorded and quantified, were fundamental to the development of Newtonian mechanics. Kepler's laws of planetary motion, derived from painstaking analysis of Tycho Brahe's extensive astronomical data, provided a mathematical description of planetary orbits, replacing earlier, less precise models. The ability to quantify the positions and movements of celestial bodies allowed for the formulation of precise mathematical laws governing their behavior, a testament to the growing power of counting in scientific discovery. Further development of astronomical instruments, improved accuracy in measurements, and the use of statistical techniques for data analysis continually refined our understanding of the cosmos. Each refinement, each incremental improvement in accuracy hinged upon the precision of counting, a seemingly simple act that underpinned revolutionary changes in understanding.

Similarly, the rise of physics during the Enlightenment was inseparable from the application of quantitative methods. Newton's laws of motion and universal gravitation, cornerstones of classical mechanics, relied heavily on mathematical formulations and precise measurements. The ability to express physical phenomena in terms of mathematical equations allowed for the prediction of future behavior, testing the validity of the proposed laws against empirical data. The counting of variables, the precise measurement of forces and accelerations, and the application of calculus to dynamic systems all contributed to the remarkable achievements in Newtonian physics. This integration of mathematics and physics established a powerful paradigm for scientific inquiry, demonstrating the indispensable role of quantitative analysis in understanding the physical world. Indeed, the success of Newtonian physics served as a powerful example for other scientific disciplines, inspiring the adoption of quantitative methods across a wider range of scientific pursuits.

The biological sciences also experienced a transformation influenced by the growing importance of counting. While earlier biological studies often relied on qualitative descriptions of organisms, the Enlightenment witnessed a gradual shift towards quantitative analysis. The development of taxonomy, for instance, involved classifying and categorizing organisms based on observable characteristics, a process that inherently involves counting and enumeration. Linnaeus's system of binomial nomenclature, a crucial advancement in biological classification, depended on the systematic counting and ordering of species, highlighting the increasingly quantitative nature of biological inquiry.

Moreover, the growing interest in human populations during the Enlightenment stimulated the development of demographic studies. The systematic collection and analysis of population data, including birth rates, death rates, and migration patterns, provided valuable insights into social trends and health conditions. Counting individuals, families, and communities, and meticulously recording demographic information, became crucial to understanding population dynamics and informed policy decisions, solidifying the importance of counting to societal well-being. This marked a significant departure from earlier, more qualitative approaches to understanding human populations, providing a powerful example of counting's capacity to reveal meaningful patterns and inform decision-making.

The impact of counting extended beyond the specific scientific disciplines. The development of new instruments and technologies during the Enlightenment further emphasized the importance of quantitative methods. The invention of the barometer and thermometer, for example, provided precise instruments for measuring atmospheric pressure and temperature. These instruments, in turn, generated quantifiable data that allowed for a deeper understanding of meteorological phenomena, and opened new avenues of scientific inquiry that depended heavily on the precision and accuracy of measurements. The development of more sophisticated microscopes, improved telescopes, and other scientific instruments provided the tools needed for increasingly precise measurement, driving forward the collection of quantitative data across a broad range of scientific domains.

The increasing reliance on quantitative methods during the Enlightenment also highlighted the reciprocal relationship between science and mathematics. Scientific advancements often stimulated the development of new mathematical techniques, while mathematical innovations provided the tools for making further scientific discoveries. The development of calculus, for example, was spurred by the need to solve problems in physics, while calculus, in turn, became an indispensable tool for making further progress in physics and other sciences. This symbiotic relationship fostered a continuous cycle of innovation, where progress in one field fueled advancements in the other.

The Enlightenment emphasis on reason and precision fostered a culture of careful data collection and rigorous analysis. Scientists began to understand the importance of controlling variables, designing experiments to minimize bias, and employing statistical methods to analyze data effectively. This increased rigor in scientific methodology, driven by a heightened appreciation for the importance of counting, led to a significant increase in the reliability and validity of scientific findings. The increasing use of statistical methods in the analysis of data further solidified this shift toward quantitative analysis, allowing scientists to identify patterns, make inferences, and draw more robust conclusions from their observations. The ability to quantify uncertainty and to estimate the likelihood of various outcomes added a new level of sophistication to scientific investigations.

However, this increased focus on quantitative methods was not without its limitations. Some critics argued that the emphasis on quantification neglected the qualitative aspects of phenomena, potentially overlooking important nuances and complexities. The reduction of complex systems into quantifiable variables could, in some cases, lead to oversimplification and an incomplete understanding of the phenomena being studied. Nevertheless, the impact of counting on scientific knowledge during the Enlightenment was profound and undeniable, transforming the nature of scientific inquiry and accelerating the pace of scientific discovery. The legacy of the Enlightenment's emphasis on quantitative methods continues to shape scientific practice today, demonstrating the enduring power of counting in our quest to understand the world around us. The pursuit of precision, driven by the Age of Reason, transformed science, building a foundation upon which much of modern scientific understanding rests. The seemingly simple act of counting became a powerful catalyst for progress, shaping our understanding of the universe and our place within it.

Chapter 7: The 19th and 20th Centuries: Counting and the Rise of Modern Mathematics

Non-Euclidean Geometries

The Enlightenment's emphasis on rigorous, quantitative methods laid the groundwork for a profound revolution in geometry during the 19th century. For millennia, Euclidean geometry, based on Euclid's axioms and postulates, had reigned supreme, providing the seemingly unquestionable framework for understanding space. However, the seeds of doubt were sown as mathematicians grappled with the seemingly simple, yet ultimately intractable, parallel postulate. This postulate, stating that through a point not on a given line, only one line can be drawn parallel to the given line, felt less self-evident than Euclid's other axioms. Attempts to prove it from the other axioms, undertaken by generations of mathematicians, all failed. This persistent failure hinted at the possibility of geometries that did not conform to Euclidean axioms.

One of the key figures in this intellectual upheaval was Carl Friedrich Gauss, a mathematical prodigy whose contributions spanned numerous fields. Gauss, a master of both theoretical and applied mathematics, privately explored the possibility of non-Euclidean geometries, meticulously investigating the implications of rejecting the parallel postulate. His notebooks reveal years of intensive work on geometries where the parallel postulate did not hold, exploring concepts that contradicted traditional Euclidean notions of space. However, Gauss, perhaps fearing the potential controversy and rejection from the established mathematical community, refrained from publishing his findings. His reservations, though understandable given the revolutionary nature of his work, delayed the wider acceptance of non-Euclidean geometries.

Independently, and unaware of Gauss's work, János Bolyai, a Hungarian mathematician, and Nikolai Ivanovich Lobachevsky, a Russian mathematician, arrived at similar conclusions. Bolyai, encouraged by his father, a prominent mathematician himself, examined the problem of the parallel postulate, eventually developing a consistent geometry where multiple parallel lines could be drawn through a point not on a given line. This geometry, known as hyperbolic geometry, challenged the very foundation of Euclidean geometry, demonstrating that alternative, logically consistent geometric systems could exist. Bolyai's work, published in 1832 as an appendix to his father's book, was a landmark achievement, though its revolutionary significance was initially slow to be recognized.

Similarly, Lobachevsky, also working independently, formulated a geometry strikingly similar to Bolyai's, publishing his findings between 1829 and 1830. His work, initially met with skepticism and even ridicule, systematically explored the consequences of rejecting the parallel postulate, developing a rigorous system of hyperbolic geometry. Despite the initial resistance, Lobachevsky's persistence and the inherent consistency of his system gradually gained him recognition, further solidifying the acceptance of non-Euclidean geometries within the mathematical community.

The implications of non-Euclidean geometries were far-reaching, extending beyond the purely theoretical realm. The existence of multiple consistent geometries shattered the long-held belief that Euclidean geometry was the unique, absolute description of space. It opened up a new perspective on the nature of mathematical truth, demonstrating that mathematical systems could be consistent even if they contradicted intuitive notions of space. This broadened the scope of mathematics, paving the way for more abstract and axiomatic approaches to mathematical inquiry.

One important consequence was the shift in emphasis from the intuitive understanding of space to a more rigorous, axiomatic approach. Instead of relying on intuitive notions of what space "should" be like, mathematicians began to focus on the logical consistency of axiomatic systems. The development of non-Euclidean geometries demonstrated that the consistency of a mathematical system did not depend on its correspondence to our perceived reality. This realization profoundly impacted the development of modern mathematics, shifting the focus toward abstract systems and their logical properties rather than their intuitive representation.

Furthermore, non-Euclidean geometries found unexpected applications in other branches of mathematics and science. Riemannian geometry, a generalization of non-Euclidean geometries developed by Bernhard Riemann, provided a framework for understanding curved spaces, which proved crucial in Einstein's theory of general relativity. Einstein's theory describes gravity not as a force but as a curvature of spacetime, a concept that relies heavily on Riemannian geometry's description of non-Euclidean spaces. The curvature of spacetime, a fundamental concept in general relativity, could not be adequately described using Euclidean geometry. The development of non-Euclidean geometries, therefore, proved essential for formulating a consistent theory of gravity, a validation of the power of abstract mathematical concepts to reshape our understanding of the physical world.

The applications of non-Euclidean geometries extended beyond physics. In cartography, the representation of the Earth's curved surface on a flat map inevitably involves distortions. Different map projections use various geometric principles to minimize different types of distortion. Understanding the properties of non-Euclidean geometries proved crucial in developing accurate and useful map projections that minimized distortion, a practical application highlighting the utility of abstract mathematical concepts.

The development of non-Euclidean geometries also impacted the philosophy of mathematics. The existence of multiple consistent geometries challenged the Kantian view that geometry was an a priori, synthetic truth, inherent to human understanding. The discovery that alternative consistent geometries could exist demonstrated that geometrical truths were not absolute and self-evident but rather depended on the chosen axiomatic system. This shift had profound implications for the epistemology of mathematics, influencing the development of various schools of mathematical thought.

In summary, the development of non-Euclidean geometries represents a pivotal moment in the history of mathematics. The courage to challenge the established paradigm, the meticulous investigation of the implications of rejecting the parallel postulate, and the eventual acceptance of alternative geometric systems significantly broadened the scope of mathematical inquiry. The impact of this development resonates through various branches of mathematics and science, highlighting the power of abstract mathematical ideas to reshape our understanding of the world and ourselves. The seemingly simple act of questioning a fundamental assumption led to a profound revolution, demonstrating the transformative potential of mathematical innovation and its enduring influence on scientific progress. The legacy of Gauss, Bolyai, and Lobachevsky extends far beyond their individual contributions; their work fundamentally altered the course of mathematics, leaving an indelible mark on our understanding of geometry and its applications in the physical sciences and beyond. The story of non-Euclidean geometry is not simply a story about lines and angles; it is a powerful testament to the human capacity for intellectual curiosity, the relentless pursuit of truth, and the transformative power of mathematical thought. It demonstrates that even the most deeply ingrained assumptions can be challenged, and that by doing so, we can unlock entirely new perspectives on the nature of reality itself. The journey from the seemingly self-evident axioms of Euclid to the complex and fascinating world of non-Euclidean geometries is evidence of the dynamic and ever-evolving nature of mathematical understanding, a journey that continues to shape our understanding of the universe and our place within it.

Set Theory and the Foundations of Mathematics

The late 19th and early 20th centuries witnessed a profound re-evaluation of the foundations of mathematics, fueled largely by the expanding field of set theory. Before this period, mathematical concepts were often treated intuitively, with a reliance on geometric intuition or informal reasoning. The rise of set theory, however, introduced a new level of rigor and abstraction, forcing mathematicians to confront fundamental questions about the nature of numbers, infinity, and the very nature of mathematical truth. This revolution was spearheaded by Georg Cantor, a German mathematician whose work on infinite sets profoundly altered the landscape of mathematics.

Cantor's initial interest lay in the study of trigonometric series, a seemingly unrelated area. His investigations led him to grapple with the concept of the cardinality of sets – essentially, the size of a set, regardless of the nature of its elements. For finite sets, the concept of cardinality is straightforward: we simply count the elements. However, Cantor's genius lay in extending this concept to infinite sets. He showed that, contrary to intuition, not all infinite sets are the same size. This revolutionary idea was encapsulated in his demonstration that the set of real numbers is "larger" than the set of natural numbers (1, 2, 3, ...), even though both are infinite. This profound result overturned long-held assumptions about infinity and opened up a completely new realm of mathematical investigation.

Cantor's proof employed a technique known as diagonalization, a clever argument showing that any attempt to establish a one-to-one correspondence between the real numbers and the natural numbers would necessarily fail. This seemingly simple yet powerful technique had far-reaching implications, demonstrating that there are different "sizes" of infinity, an idea that challenged the very notion of infinity itself. He introduced the concept of transfinite numbers, extending the familiar number system beyond the realm of finite numbers to encompass infinite quantities. He denoted the cardinality of the natural numbers as $\aleph_0$ (aleph-null), the smallest infinite cardinal number. The cardinality of the real numbers, he showed, is a larger infinity, often denoted as **c** (for continuum). The existence of these different infinities challenged long-held beliefs about the nature of infinity and opened up new avenues of research in mathematics.

The introduction of transfinite numbers was not without controversy. Many mathematicians, deeply entrenched in the established framework of mathematics, were deeply skeptical of Cantor's work. His ideas, particularly his treatment of infinity, were considered radical and even paradoxical by some. Critics such as Leopold Kronecker famously argued against Cantor's work, claiming that it dealt with mathematical objects that did not possess concrete existence. This resistance to Cantor's ideas highlights the conservative nature of the mathematical community and the inherent difficulties involved in accepting fundamental shifts in mathematical understanding. Despite this resistance, Cantor's work gradually gained acceptance as its internal consistency and power became increasingly apparent.

Cantor's set theory provided a new language and framework for expressing and analyzing mathematical concepts. Many seemingly disparate mathematical areas found a unifying framework within set theory, leading to a deeper understanding of their underlying structures. The formalization of mathematical concepts in terms of sets offered a new level of rigor, reducing the reliance on intuitive arguments and enhancing the precision of mathematical reasoning. This increased rigor had a profound impact on the development of mathematics, paving the way for a more axiomatic approach to mathematical foundations.

However, the very success of set theory in providing a foundation for mathematics also brought to light some disturbing paradoxes. These paradoxes, which stemmed from seemingly simple and intuitive set-theoretic operations, highlighted the need for a more careful and axiomatic treatment of set theory. One of the most famous of these was Russell's Paradox, discovered by Bertrand Russell in 1901. Russell's paradox involved the concept of a "set of all sets that do not contain themselves." This seemingly innocuous definition leads to a contradiction, illustrating the limitations of unrestricted set formation. If this set contains itself, it violates its own definition; if it doesn't contain itself, it again violates its definition. This paradox exposed a fundamental flaw in the naive understanding of set theory, highlighting the need for a more carefully defined axiomatic system.

The discovery of paradoxes such as Russell's spurred the development of axiomatic set theories. These systems aimed to formalize the basic axioms and rules of set theory in a way that avoided the paradoxes that had plagued earlier, more naive approaches. One of the most influential axiomatic set theories is Zermelo-Fraenkel set theory (ZFC), which incorporates the axiom of choice. This theory provides a rigorous framework for constructing and manipulating sets, while simultaneously preventing the occurrence of known paradoxes. The axiom of choice, in particular, while seemingly innocuous, is a source of both mathematical power and philosophical debate, with its implications stretching into various areas of mathematics.

The development of axiomatic set theories was a pivotal moment in the history of mathematics. It marked a shift from a largely intuitive and informal approach to a more rigorous and formal one. This increased emphasis on formal systems and axioms had a profound impact on the philosophy of mathematics, leading to various schools of thought regarding the nature of mathematical truth and the foundations of mathematics. Formalism, for example, viewed mathematics as a purely formal system, independent of any interpretation or meaning. Logicism, on the other hand, sought to reduce mathematics to logic, treating mathematical truths as logical consequences of fundamental logical principles. Intuitionism, a more radical approach, questioned the acceptance of the law of excluded middle and the existence of non-constructively defined objects. The ongoing debate concerning the foundations of mathematics reflects the profound and far-reaching consequences of the set-theoretic revolution.

The influence of set theory extends far beyond the foundations of mathematics. It has become a fundamental tool in many branches of mathematics, providing a unifying framework for diverse mathematical concepts. Topology, analysis, algebra, and even parts of physics utilize set theory as a fundamental language and tool. The power and generality of set-theoretic methods have led to the development of sophisticated mathematical tools used across a broad spectrum of disciplines.

In summary, the development of set theory in the 19th and 20th centuries marked a profound turning point in the history of mathematics. Cantor's groundbreaking work on infinite sets, though initially met with skepticism, revolutionized our understanding of infinity and paved the way for a more rigorous and abstract approach to mathematics. The subsequent discovery of paradoxes, such as Russell's paradox, led to the development of axiomatic set theories, which provided a more robust foundation for mathematics. The implications of set theory extend beyond mathematical foundations, impacting the philosophy of mathematics and finding widespread applications in various branches of the discipline. The story of set theory illustrates the often-turbulent evolution of mathematical thought, characterized by both revolutionary breakthroughs and the inevitable struggle to reconcile new ideas with established paradigms. The enduring legacy of Cantor and the subsequent contributors to the field underscores the dynamic and ever-evolving nature of mathematical knowledge and its continuing influence on our understanding of the world. The journey from naive set theory to the rigorous axiomatic systems of today is validation of persistent human pursuit of understanding and the enduring power of mathematical reasoning.

Abstract Algebra and Group Theory

The late 19th and early 20th centuries witnessed not only a revolution in the foundations of mathematics through set theory, but also a parallel surge in abstract algebra, particularly group theory. While seemingly disparate at first glance, both movements shared a common thread: a shift away from concrete examples and towards abstract structures and general principles. This move toward abstraction, fueled by the need to solve seemingly intractable problems, led to the development of powerful mathematical tools with wide-ranging applications.

Before this period, algebra was largely concerned with solving equations, often focusing on specific types of equations and their solutions. The development of abstract algebra marked a significant departure from this approach. Instead of focusing on specific equations, abstract algebra focused on the underlying structures and relationships between mathematical objects. This shift was driven by the realization that many different mathematical objects could exhibit similar structural properties, leading to the concept of isomorphic structures: structures that are essentially the same despite superficial differences in their elements.

One of the crucial milestones in the development of abstract algebra was the work of Évariste Galois, a brilliant but tragically short-lived French mathematician. Galois, in his groundbreaking work on the solvability of polynomial equations, introduced the concept of a group, which would become a cornerstone of abstract algebra. His work was far ahead of its time, and its full significance was not fully appreciated until much later.

Galois's initial focus was on determining which polynomial equations could be solved using radicals, that is, using a finite sequence of additions, subtractions, multiplications, divisions, and the extraction of roots. Before Galois, mathematicians had known how to solve quadratic, cubic, and quartic equations using radicals. However, the problem of finding a general method for solving polynomial equations of degree five or higher remained unsolved. Galois tackled this problem by associating a group with each polynomial equation. The group, in essence, captured the symmetries of the roots of the equation. He demonstrated that the solvability of a polynomial equation by radicals was intimately connected to the properties of its associated group. Specifically, he showed that a polynomial equation is solvable by radicals if and only if its Galois group is a solvable group, a group with a specific hierarchical structure. This stunning result provided a complete characterization of polynomial equations solvable by radicals, settling a question that had baffled mathematicians for centuries.

The concept of a group, as introduced by Galois, is remarkably simple in its fundamental definition. A group is a set equipped with a binary operation, a way of combining two elements of the set, that satisfies four specific axioms: *closure* (the result of combining two elements is always in the set), *associativity* (the order of operations doesn't matter), *the existence of an identity element* (an element that leaves other elements unchanged when combined), and *the existence of inverses* (for every element, there is another element that, when combined, yields the identity element). Despite this simple definition, groups exhibit a surprising richness of structure and have far-reaching applications in various areas of mathematics and science.

Consider, for instance, the set of integers with the operation of addition. This set forms a group, with 0 being the identity element and -n being the inverse of n. Similarly, the set of non-zero real numbers forms a group under multiplication, with 1 as the identity and 1/n as the inverse of n. These are examples of infinite groups. However, there are also finite groups, such as the group of rotations of a regular polygon. The symmetries of a regular polygon, the rotations that leave the polygon unchanged, form a finite group. The order of a group is simply the number of elements it contains, and this provides a link back to counting, albeit in a significantly more abstract setting.

Galois's work, while revolutionary, remained largely unknown for several decades. However, it eventually became a central pillar in the development of modern abstract algebra. Mathematicians began to study groups in their own right, examining their structure, properties, and relationships. This led to the development of a rich body of theory, including the classification of finite simple groups, one of the most significant achievements in 20th-century mathematics. This classification theorem states that every finite simple group belongs to one of a few well-understood families of groups, which is a monumental achievement in understanding the structure of finite groups. The sheer magnitude of the proof involved thousands of pages written by hundreds of mathematicians over several decades.

The development of abstract algebra extended beyond group theory. Other algebraic structures, such as rings and fields, were also extensively studied. Rings are sets with two operations, typically addition and multiplication, satisfying certain axioms, while fields are rings with additional properties, including the existence of multiplicative inverses for all non-zero elements. These structures arise naturally in many contexts, from number theory to linear algebra. For example, the integers form a ring, while the real numbers form a field. These structures provide frameworks for understanding arithmetic operations and their properties in a more abstract and general way.

The impact of abstract algebra is felt across numerous branches of mathematics and beyond. It provides essential tools for studying symmetries, classifying mathematical objects, and solving complex problems. Its applications range from cryptography and coding theory to physics and chemistry. Group theory, in particular, plays a vital role in understanding the symmetries of physical systems, leading to a deeper understanding of the fundamental laws of nature.

The abstraction inherent in abstract algebra allowed for a unification of different mathematical areas. Concepts that seemed unrelated at first glance, such as the rotations of a polygon and the solutions to polynomial equations, were found to have underlying similarities that could be understood through the lens of abstract algebraic structures. This unification provides a powerful framework for making connections between seemingly disparate mathematical ideas and concepts, thereby expanding our understanding of the mathematical universe.

Moreover, the development of abstract algebra underscores the powerful role of abstraction in mathematical progress. By moving away from specific examples and focusing on underlying structures, mathematicians were able to develop powerful and widely applicable tools, revealing deep connections and unifying seemingly disparate mathematical fields. The legacy of Galois and the subsequent development of abstract algebra exemplify the power of mathematical abstraction and its capacity to reveal profound insights into the nature of mathematical structures and their relationship to the world around us. It illustrates that counting, at its core, is a foundation not only for arithmetic, but for the very structure of mathematics itself, a legacy that continues to unfold even today. The continued study of abstract algebra promises further discoveries and advancements in understanding the intricate structures underpinning mathematics and its applications.

The Rise of Computer Science and its Dependence on Counting

The seemingly abstract world of 19th-century mathematics, particularly abstract algebra, found an unexpectedly concrete and impactful application in the expanding field of computer science during the 20th century. The very architecture of the digital computer, the backbone of the information age, rests fundamentally upon principles of counting and discrete mathematics, as validated in the enduring power and universality of mathematical concepts. The seemingly simple act of counting, refined and formalized over millennia, became the cornerstone of a technological revolution.

The development of the digital computer was not a sudden leap but rather the culmination of decades of theoretical and engineering advancements. Crucially, the binary system, a counting system based on only two digits, **0** and **1**, proved perfectly suited to the needs of electronic computation. Unlike the decimal system we use in everyday life, the binary system's inherent simplicity aligned remarkably well with the "on" and "off" states of electronic switches, transistors, and other components within a computer. Each "on" state could be represented by a **1**, and each "off" state by a **0**. This straightforward mapping between mathematical symbols and physical states forms the very foundation of digital computation. This seemingly simplistic choice has profound implications, forming the language through which computers process information.

The elegance and efficiency of the binary system are directly linked to its mathematical underpinnings. Binary numbers, sequences of **0**s and **1**s, can represent any integer, and with the addition of a radix point, any real number. The arithmetic operations of addition, subtraction, multiplication, and division can all be performed within the binary system, employing simple logic gates that manipulate these **0**s and **1**s. These gates, the fundamental building blocks of digital circuits, are essentially physical manifestations of Boolean algebra, a branch of algebra dealing with logical operations. Boolean algebra, developed by George Boole in the mid-19th century, provided the theoretical framework for designing and analyzing logical circuits.

Boolean algebra operates on a set of only two values, typically represented as true (**1**) and false (**0**). The fundamental operations are NOT (inversion), AND (conjunction), and OR (disjunction). These operations, remarkably simple in their definition, can be combined to perform complex logical operations. For instance, a NOT gate inverts an input signal; if the input is **1** (true), the output is **0** (false), and vice versa. An AND gate outputs **1** only if both of its inputs are **1**, otherwise it outputs **0**. An OR gate outputs **1** if *one* or more of its inputs are **1**, otherwise it outputs **0**. These simple operations, when combined in intricate circuits, can perform incredibly complex calculations.

The connection between Boolean algebra and the design of digital computers is profound. Every logical operation performed by a computer, from comparing numbers to executing program instructions, ultimately boils down to a sequence of these fundamental Boolean operations executed by the underlying logic gates. The design of complex circuits, such as adders, multipliers, and memory units, relies heavily on the principles of Boolean algebra to ensure that they correctly perform their intended function. The minimization of logic gates in a circuit, an important aspect of efficient circuit design, often involves sophisticated mathematical techniques rooted in Boolean algebra and discrete mathematics.

Furthermore, the way data is represented and manipulated within a computer is intrinsically linked to counting and discrete mathematics. Data structures, such as arrays, linked lists, trees, and graphs, are all mathematical constructs that play crucial roles in organizing and accessing data efficiently. Algorithms, the step-by-step procedures that computers use to solve problems, are also deeply rooted in discrete mathematics. The analysis of algorithms, including their efficiency and complexity, relies heavily on mathematical techniques, such as asymptotic notation (Big O notation), which describes how the runtime or memory usage of an algorithm scales with the size of the input.

The development of programming languages, the means through which humans interact with computers, further demonstrates the intertwining of counting and computation. Programming languages are built upon formal systems of logic and syntax, based on precise mathematical rules. The compilation or interpretation of code involves complex processes that rely on techniques from formal language theory and automata theory, both branches of theoretical computer science with strong ties to discrete mathematics.

The impact extends beyond the fundamental architecture and design of computers. Areas such as cryptography, the science of secure communication, rely heavily on number theory and abstract algebra. Public-key cryptography, widely used to secure online transactions, utilizes advanced mathematical concepts, such as modular arithmetic and prime factorization, to ensure confidentiality and data integrity. These cryptographic techniques are a sophisticated application of counting and abstract mathematical structures, protecting sensitive information in the digital realm.

The influence of counting continues to grow as computer science advances into new frontiers. Quantum computing, for instance, leverages the principles of quantum mechanics to perform computations in ways that are fundamentally different from classical computers. While still in its emerging stages, quantum computing already presents fascinating mathematical challenges and opportunities, relying on principles of linear algebra and quantum information theory, highlighting the ongoing interplay between mathematics and computational technologies.

In conclusion, the rise of computer science is inextricably linked to the history of counting. From the binary representation of data to the sophisticated algorithms that govern computation, discrete mathematics and, more broadly, the abstract structures of mathematics are the very foundations upon which the digital world is built. The seemingly simple act of counting, explored throughout this book from its origins in prehistoric tally marks to its modern applications, has played an essential role in shaping the technology that defines the modern era. The journey from counting on fingers to the complexities of modern computing is evident in the power and universality of mathematics and its profound impact on human civilization. The ongoing evolution of computer science continues to unveil new and unexpected connections between the abstract realm of mathematics and the concrete realities of our increasingly digital world. The story of counting is far from over; it is a continuing narrative that will undoubtedly shape the technological landscape of tomorrow.

The Impact of Modern Mathematics on Science and Technology

The remarkable progress in science and technology throughout the 19th and 20th centuries is inextricably linked to the parallel advancements in mathematics. No longer a mere tool for calculation, mathematics evolved into a powerful engine driving innovation, providing the theoretical frameworks and analytical tools necessary to understand and manipulate the natural world. The development of sophisticated mathematical models, far removed from the simple counting systems of antiquity, became essential for advancements across a wide range of disciplines.

One of the most profound examples of this synergy between mathematics and scientific progress lies in the field of physics. Classical mechanics, the study of motion and forces, underwent a radical transformation with the advent of calculus, developed independently by Isaac Newton and Gottfried Wilhelm Leibniz. Newton's laws of motion, expressed in the elegant language of calculus, provided a precise mathematical framework for describing the motion of objects, from projectiles to planets. The ability to predict planetary orbits with unprecedented accuracy, a testament to the power of calculus, cemented its position as an indispensable tool for scientific inquiry.

The rise of electromagnetism in the 19th century further highlighted the symbiotic relationship between mathematics and physics. James Clerk Maxwell's equations, a set of four elegant equations describing the behavior of electric and magnetic fields, are a stunning example of the power of mathematical formalism in unifying seemingly disparate phenomena. Maxwell's equations, based on vector calculus and differential equations, not only predicted the existence of electromagnetic waves but also laid the foundation for the development of radio, television, and countless other technologies that rely on electromagnetic radiation. The precision and predictive power of Maxwell's equations underscore the importance of rigorous mathematical reasoning in unraveling the complexities of the physical world.

The development of relativity in the early 20th century represents another watershed moment in the history of science, deeply intertwined with mathematical innovation. Albert Einstein's theories of special and general relativity revolutionized our understanding of space, time, gravity, and the universe itself. These theories, expressed in the language of advanced calculus, tensor calculus, and differential geometry, profoundly altered the course of physics and cosmology. The mathematical sophistication of relativity was not merely an aesthetic choice; it was essential for formulating and testing the theory, leading to predictions that have been repeatedly confirmed by experimental observation.

The advent of quantum mechanics in the early 20th century presented even more profound mathematical challenges. The probabilistic nature of quantum phenomena necessitated the development of new mathematical tools, including linear algebra, Hilbert spaces, and operator theory. The mathematical formalism of quantum mechanics, while often abstract and counterintuitive, provides an incredibly precise and accurate description of the behavior of matter at the atomic and subatomic levels. This mathematical framework has underpinned numerous technological advancements, from lasers and transistors to nuclear energy and medical imaging, showcasing the transformative power of mathematical innovation in shaping modern technology.

Beyond physics, the influence of modern mathematics extends across a vast range of scientific and technological disciplines. In engineering, mathematical models are crucial for designing and analyzing structures, circuits, and systems. Differential equations, for instance, are essential for modeling the behavior of mechanical systems, electrical circuits, and fluid flows. Numerical methods, algorithms for solving mathematical problems using computers, have become indispensable tools for engineers, allowing them to simulate complex systems and optimize their designs. The development of powerful computational methods has enabled engineers to design increasingly sophisticated and efficient structures, devices, and systems.

In the biological sciences, mathematical modeling has played a crucial role in understanding the complexities of biological systems. Mathematical models are used to study population dynamics, the spread of diseases, the evolution of species, and the functioning of ecological systems. The development of sophisticated statistical techniques has allowed biologists to analyze vast datasets, identify patterns, and draw conclusions that would have been impossible without the tools of modern mathematics. Mathematical biology is a rapidly growing field, highlighting the increasing importance of mathematical reasoning in understanding the intricacies of life.

The field of computer science, as previously discussed, is inherently mathematical. The architecture of computers, the design of algorithms, the development of programming languages, and the analysis of data all rely heavily on mathematical foundations. Discrete mathematics, including graph theory, combinatorics, and number theory, plays a vital role in numerous areas of computer science, from data structures and algorithms to cryptography and network security. The development of sophisticated mathematical algorithms has enabled the creation of powerful computing systems that can process and analyze vast amounts of data, transforming numerous aspects of modern life.

The development of sophisticated statistical methods in the 20th century has also had a transformative impact on various scientific disciplines. Statistical methods are used to analyze experimental data, identify patterns, draw inferences, and test hypotheses. The development of statistical techniques, such as regression analysis, hypothesis testing, and Bayesian inference, has enabled scientists to make more robust and reliable conclusions from their data. The increasing availability of large datasets, coupled with the development of powerful computational tools, has led to an explosion in the use of statistical methods across a vast range of scientific fields, from medicine and genetics to economics and climate science.

The ongoing development of new mathematical theories and techniques continues to drive innovation in science and technology. For example, the field of chaos theory, which studies the behavior of complex nonlinear systems, has provided new insights into a wide range of phenomena, from weather patterns to the human heart. The application of fractal geometry, the study of irregular shapes and patterns, has led to new approaches in image compression, signal processing, and other areas. The development of new mathematical tools, such as machine learning algorithms, is revolutionizing fields such as artificial intelligence and data science, driving significant technological advancements and opening up new possibilities for scientific discovery.

In conclusion, the intertwined history of mathematics and science is a tribute to the power of human ingenuity. The seemingly abstract world of mathematics has proven to be an incredibly effective tool for understanding the physical world, driving scientific discovery and technological innovation. From the precise predictions of planetary orbits to the intricate workings of quantum mechanics, mathematics has provided the theoretical frameworks and analytical tools that have enabled humanity to make unprecedented progress in understanding and manipulating the universe. The ongoing interplay between mathematics and science promises continued advancements in both fields, pushing the boundaries of human knowledge and shaping the technological landscape of the future. The legacy of counting, meticulously documented from its rudimentary origins, is, therefore, not merely a story of numbers; it's a story of human progress, a narrative etched in the very fabric of scientific and technological advancement. The future will undoubtedly see further integration, leading to exciting new discoveries and innovations.

Chapter 8: Counting in the 21st Century: From Quantum Physics to Artificial Intelligence

Quantum Mechanics and its Mathematical Foundations

The remarkable journey of counting, from the rudimentary tally marks of our ancestors to the sophisticated algorithms driving today's computers, culminates in a profound encounter with the quantum realm. Quantum mechanics, a theory governing the behavior of matter at the atomic and subatomic levels, validates the power of abstract mathematics to illuminate the deepest mysteries of the universe. Its development, however, was not a straightforward path; it demanded the creation of entirely new mathematical tools, far removed from the simple arithmetic of ancient civilizations. The transition from classical mechanics to quantum mechanics involved a paradigm shift, not just in our understanding of the physical world, but also in the mathematical language we used to describe it.

Classical mechanics, as we saw, found its mathematical expression in the elegant calculus of Newton and Leibniz. This system, based on deterministic laws, allowed for precise prediction of the motion of objects given their initial conditions. The trajectory of a projectile or the orbit of a planet could be calculated with remarkable accuracy. However, the microscopic world revealed itself to be far more unruly, less predictable. The behavior of electrons orbiting an atom, for example, could not be accurately described using the deterministic framework of classical mechanics. This required a radical rethinking of the very foundations of physics, leading to the birth of quantum mechanics in the early 20th century.

The mathematical foundation of quantum mechanics departs significantly from that of classical mechanics. While classical mechanics relies heavily on real numbers to represent physical quantities such as position, momentum, and energy, quantum mechanics utilizes the far richer and more abstract world of complex numbers. A complex number is a number that can be expressed in the form a + bi, where 'a' and 'b' are real numbers, and 'i' is the imaginary unit, defined as the square root of -1. The use of complex numbers might seem initially arbitrary, a purely mathematical contrivance. Yet, it turns out to be essential for capturing the wave-like nature of quantum particles.

The wave function, a central concept in quantum mechanics, is a complex-valued function that describes the state of a quantum system. It is not simply a mathematical abstraction; it encodes all the information we can possibly know about the system. The square of the magnitude of the wave function gives the probability density of finding the particle at a particular location. This inherent probabilistic nature stands in stark contrast to the deterministic predictions of classical mechanics. In classical mechanics, we can precisely determine the position and momentum of a particle at any given time. In quantum mechanics, we can only talk about the probabilities of finding the particle in different states.

The formalism of quantum mechanics relies heavily on linear algebra, a branch of mathematics dealing with vectors, matrices, and linear transformations. The state of a quantum system is represented by a vector in a Hilbert space, an infinite-dimensional vector space with specific mathematical properties. Quantum mechanical operators, which represent physical observables like position, momentum, and energy, are linear operators acting on these vectors. The act of measurement in quantum mechanics corresponds to applying these operators to the state vector. The result of the measurement is one of the eigenvalues of the operator, and the state vector collapses to the corresponding eigenvector. The intricate dance of vectors, matrices, and operators within the Hilbert space forms the mathematical backbone of quantum mechanics, allowing physicists to predict the outcomes of experiments with remarkable accuracy.

The use of linear algebra is not merely a convenient mathematical framework; it is inextricably linked to the fundamental properties of quantum systems. The principle of superposition, for example, states that a quantum system can exist in a superposition of multiple states simultaneously. This concept finds its natural mathematical expression in the linear combination of state vectors in the Hilbert space. Similarly, the phenomenon of entanglement, where two or more quantum particles become linked in such a way that their fates are intertwined, regardless of the distance separating them, is elegantly described using the tensor product of Hilbert spaces.

One of the most striking implications of quantum mechanics is its inherent uncertainty. The Heisenberg uncertainty principle states that it is impossible to simultaneously measure the position and momentum of a particle with arbitrary precision. The product of the uncertainties in position and momentum must be greater than or equal to a fundamental constant, Planck's constant. This uncertainty is not a limitation of our measurement instruments; it is a fundamental property of the quantum world, reflecting the inherent fuzziness of quantum states. This uncertainty principle, too, has a precise mathematical formulation within the framework of quantum mechanics.

The development of quantum computation further highlights the profound interplay between mathematics and physics. Quantum computers harness the principles of quantum mechanics to perform computations that are intractable for even the most powerful classical computers. Quantum algorithms, such as Shor's algorithm for factoring large numbers and Grover's algorithm for searching unsorted databases, leverage the principles of superposition and entanglement to achieve exponential speedups over classical algorithms. The design and analysis of quantum algorithms require a deep understanding of both quantum mechanics and the underlying mathematical structures.

Quantum computing also necessitates a sophisticated understanding of error correction and fault tolerance. Quantum systems are inherently fragile, susceptible to decoherence—the loss of quantum information due to interactions with the environment. Developing robust quantum computers requires the development of sophisticated error correction codes, which are based on advanced mathematical concepts such as quantum error-correcting codes. These codes ensure that the computation is reliable, despite the noise and errors inherent in quantum systems.

The journey from the simple counting of our ancestors to the complex mathematics of quantum mechanics is a remarkable demonstration of human ingenuity. The seemingly abstract world of complex numbers, linear algebra, and Hilbert spaces has proven essential in understanding the deepest mysteries of the universe. The precision and accuracy with which quantum mechanics describes the behavior of matter at the atomic and subatomic levels is a compelling evidence of the power of mathematical formalism in unraveling the secrets of nature. This mathematical framework is not merely an aesthetic choice; it is essential for making predictions and developing technologies that rely on quantum phenomena, from lasers and transistors to medical imaging and quantum computers. The story of counting, therefore, continues to unfold, taking us ever deeper into the heart of reality, constantly pushing the boundaries of what we can understand and achieve. The implications of this ongoing evolution, both for scientific discovery and technological innovation, are vast and yet to be fully explored. The future undoubtedly holds even more surprises as we continue to explore the mathematical underpinnings of the universe and unlock the potential of quantum technologies.

Big Data and Statistical Analysis

The explosive growth of digital information has ushered in the era of Big Data, a realm where the sheer volume, velocity, and variety of data pose unprecedented challenges and opportunities for analysis. This deluge of data, encompassing everything from social media interactions and online transactions to genomic sequences and sensor readings, demands new approaches to counting and statistical analysis. While the fundamental principles of counting remain unchanged, the meticulous tallying of events and quantities, the scale and complexity of Big Data necessitate sophisticated techniques to extract meaningful insights.

The first hurdle in tackling Big Data is its sheer size. Traditional statistical methods, designed for smaller, more manageable datasets, often prove inadequate. The computational power required to process and analyze terabytes, petabytes, or even exabytes of data dwarfs the capabilities of conventional computers. This has driven the development of distributed computing frameworks like Hadoop and Spark, which enable the parallel processing of massive datasets across clusters of machines. These frameworks break down large datasets into smaller, manageable chunks, allowing multiple processors to work concurrently, significantly reducing processing time.

The velocity of Big Data is another critical aspect. Data streams continuously flow from various sources, requiring real-time or near real-time processing. This necessitates the use of streaming algorithms, which can analyze data as it arrives, without the need to store it completely. Examples include algorithms used for fraud detection in financial transactions, real-time monitoring of power grids, or analyzing social media trends as they unfold. These algorithms need to be adaptive and robust, capable of handling noisy or incomplete data streams.

Beyond volume and velocity, the variety of Big Data presents a unique challenge. Modern datasets often comprise diverse data types, including structured data (like relational databases), semi-structured data (like XML or JSON files), and unstructured data (such as various forms of text, images, and videos). Effective analysis requires techniques capable of handling this heterogeneity. Machine learning algorithms, particularly deep learning models, have proven remarkably effective in processing unstructured data, extracting meaningful features and patterns that may not be readily apparent to human analysts. For instance, image recognition systems utilize deep learning to identify objects and patterns in images, while natural language processing techniques are employed to analyze vast quantities of text data.

The interpretation of Big Data relies heavily on statistical methods, which provide the tools for drawing inferences from data and quantifying uncertainty. However, the sheer volume and complexity of Big Data introduce new challenges for statistical inference. Traditional statistical assumptions, such as normality or independence of data points, may not hold for Big Data, requiring the development of robust and non-parametric methods. Moreover, the presence of noise, outliers, and missing data can significantly affect the reliability of results, necessitating careful data cleaning and preprocessing techniques.

One important area where Big Data and statistical analysis intersect is in the realm of causal inference. Identifying causal relationships between variables in Big Data is crucial for informed decision-making in various fields. However, establishing causality is a challenging task, as mere correlation does not imply causation. Advanced statistical methods, such as instrumental variables and regression discontinuity designs, are employed to address this challenge and rigorously assess causal effects in Big Data.

The applications of Big Data and statistical analysis are vast and far-reaching. In medicine, Big Data is used to analyze genomic data, identify disease patterns, and develop personalized treatments. Genomic sequencing generates enormous amounts of data, requiring advanced computational tools to analyze genetic variations and their association with diseases. Machine learning algorithms are used to predict disease risk, personalize medication dosages, and accelerate drug discovery. Similarly, in finance, Big Data plays a crucial role in fraud detection, risk management, and algorithmic trading. Sophisticated algorithms analyze financial transactions, identifying patterns that may indicate fraudulent activities. These algorithms also help financial institutions assess risks associated with investments and loans.

In the social sciences, Big Data has revolutionized the study of human behavior and social interactions. Social media data, for example, provides a rich source of information on people's opinions, beliefs, and behaviors. Statistical methods are used to analyze this data, identifying trends and patterns in public opinion, understanding the spread of information, and predicting social movements. Similarly, Big Data is used to analyze election results, predict voting behavior, and understand political polarization. Furthermore, Big Data techniques are used to study urban planning, analyzing traffic patterns, optimizing transportation systems, and improving urban infrastructure.

However, the use of Big Data also raises ethical considerations. Privacy concerns are paramount, as the collection and analysis of large datasets can potentially reveal sensitive personal information. The need for data anonymization and secure data storage is crucial to protect individual privacy. Moreover, biases in data can lead to discriminatory outcomes, reinforcing existing inequalities. Careful attention to data quality, bias detection, and fairness is essential to ensure ethical and responsible use of Big Data.

The development of new statistical methods and algorithms for analyzing Big Data is an ongoing process. Researchers are continuously developing more efficient, scalable, and robust techniques for handling the unique challenges presented by this new data paradigm. The integration of machine learning, statistical modeling, and distributed computing is crucial for extracting meaningful insights from Big Data. Furthermore, the development of new visualization techniques is essential for effectively communicating the findings derived from Big Data analysis to a wider audience. The future of data analysis depends on the continued advancement of these methodologies and the responsible application of Big Data to address the world's most pressing challenges.

In conclusion, the intersection of counting and statistical analysis in the context of Big Data represents a paradigm shift in how we collect, analyze, and interpret information. While the fundamental principles of counting remain the foundation, the sheer scale, velocity, and variety of Big Data necessitate the development of new computational tools and statistical methods. The applications of Big Data and its associated analytical techniques are broad and far-reaching, impacting various fields from medicine and finance to social sciences and urban planning. However, ethical considerations remain crucial, necessitating responsible data handling practices to mitigate privacy concerns and prevent bias. The future of Big Data and statistical analysis hinges on the continued innovation of methodologies and their responsible application to tackle complex global challenges. The seamless integration of advanced computing, innovative statistical techniques, and rigorous ethical frameworks will determine the true potential of this rapidly evolving field, allowing us to unlock profound insights from the ever-expanding digital landscape. The journey of counting, which began with simple tally marks, has reached new heights of complexity and sophistication, allowing us to understand and interact with the world in ways previously unimaginable.

Artificial Intelligence and Machine Learning

The preceding discussion on Big Data and its analytical challenges provides a crucial backdrop for understanding the role of counting within the accelerating field of Artificial Intelligence (AI) and Machine Learning (ML). While Big Data provides the raw material, AI and ML provide the sophisticated tools for extracting meaning and making predictions. At its core, AI and ML rely heavily on the fundamental principles of counting and sophisticated mathematical algorithms built upon those principles. These algorithms are the engines that drive the learning process, allowing machines to identify patterns, make decisions, and solve complex problems that would overwhelm human capabilities.

Consider, for example, the seemingly simple task of image recognition. A machine learning model tasked with identifying cats in images doesn't "see" a cat in the same way a human does.

Instead, it analyzes the image as a vast array of numerical data points – pixel values representing color and intensity. The core of the algorithm involves counting and comparing these numerical values against a vast dataset of labeled images (images where the presence of a cat has been confirmed). The model learns by identifying statistical regularities, counting instances of certain patterns of pixels that correlate strongly with the presence of a cat. The process refines itself iteratively, adjusting its internal parameters (weights and biases in neural networks) to minimize errors in its predictions. This iterative refinement, a process central to machine learning, is fundamentally a process of counting and adjusting counts based on feedback.

Similarly, natural language processing (NLP), a crucial branch of AI, relies heavily on counting. Understanding human language involves analyzing sequences of words, identifying grammatical structures, and extracting semantic meaning. NLP algorithms often involve counting the frequency of words, analyzing the co-occurrence of words within sentences and paragraphs, and identifying patterns in word usage. These counts provide crucial statistical information, allowing the algorithm to discern the relationships between words and understand the overall meaning of a text. For instance, sentiment analysis, determining whether a piece of text expresses positive, negative, or neutral sentiment, relies on counting the frequency of words associated with specific emotions. The more frequently "happy," "joyful," or "excited" appear, the higher the likelihood the text expresses positive sentiment.

Beyond simple counting, more complex mathematical algorithms underpin most AI and ML systems. Linear algebra, for instance, is essential in many machine learning models, particularly those involving vector spaces and matrices. Neural networks, a powerful class of machine learning models inspired by the structure of the human brain, rely heavily on matrix operations. Counting the number of connections (weights) between neurons and applying mathematical operations to these connection weights is integral to how these networks learn and make predictions. These processes, while sophisticated, are ultimately rooted in the fundamental act of counting and measuring.

Furthermore, the field of deep learning, a subfield of machine learning that uses artificial neural networks with multiple layers, heavily leverages counting and statistical analysis. Deep learning models excel at processing complex data like images, videos, and audio, extracting subtle patterns that would be difficult or impossible for humans to identify. These models learn hierarchical representations of data, where the earlier layers identify low-level features (edges, corners in images), and subsequent layers build upon these features to identify more complex patterns (objects, scenes). The learning process, again, relies on iterative adjustments of weights, guided by error minimization, which is essentially a process of counting and adjusting based on feedback.

The application of AI and ML extends to numerous fields. In medicine, AI algorithms analyze medical images (X-rays, MRIs) to detect diseases like cancer at early stages, often with greater accuracy than human experts. This process involves sophisticated counting and pattern recognition based on pixel values and other numerical data extracted from the images. In finance, AI is used for fraud detection, risk assessment, and algorithmic trading. These applications rely heavily on statistical analysis and prediction models, all rooted in fundamental counting techniques and complex mathematical relationships. In transportation, self-driving cars employ AI algorithms that process sensor data (from cameras, lidar, and radar) to navigate roads and avoid obstacles, using sophisticated counting and pattern recognition to interpret the environment.

However, the rapid advancement of AI and ML also raises significant ethical concerns. Bias in algorithms, stemming from biases in the data they are trained on, can lead to discriminatory outcomes. For instance, facial recognition systems trained primarily on images of white faces may perform poorly on individuals with darker skin tones, resulting in unfair or inaccurate outcomes. Understanding and mitigating these biases requires careful analysis of the data, including counting and analyzing the representation of different groups within datasets. Moreover, the potential for job displacement due to automation driven by AI is a major societal challenge requiring careful consideration. Addressing these ethical issues demands a critical and multidisciplinary approach, acknowledging the role of counting, algorithms, and their potential consequences.

The development and application of AI and ML represent a remarkable extension of the long history of counting. From the rudimentary tally marks of our ancestors to the complex algorithms powering modern AI, the fundamental act of counting remains the foundation upon which these sophisticated technologies are built. Understanding the mathematical foundations of AI and ML is crucial to both harnessing their immense potential and mitigating their potential risks. The journey of counting continues, reaching unprecedented levels of complexity and sophistication in the 21st century, shaping our world in profound ways. As we move forward, it is imperative that we remain mindful of the ethical implications of these technologies, ensuring that the power of counting and computation is used responsibly and equitably for the benefit of all. The future of AI and ML will be shaped not only by the ingenuity of algorithmic design but also by our collective understanding of their social and ethical implications, a challenge requiring a careful and critical examination of the entire process, from data collection to the interpretation of results, all grounded in the fundamental act of counting.

The Future of Counting Emerging Trends and Challenges

The integration of advanced computational techniques with mathematical modeling has dramatically altered our ability to count and analyze vast quantities of data, pushing the boundaries of what was previously considered feasible. This has led to unprecedented advancements in fields ranging from genomics and astrophysics to climate modeling and financial forecasting. For instance, the Human Genome Project, completed in 2003, was a monumental undertaking involving the sequencing and mapping of the entire human genome. This would have been practically impossible without the development of high-throughput sequencing technologies and sophisticated computational tools to analyze the resulting terabytes of data. Counting and comparing nucleotide sequences, identifying genes, and understanding their functions all depend heavily on complex algorithms and computational power, highlighting the profound interrelationship between advanced counting methodologies and biological understanding.

Similarly, advancements in astrophysics have been fueled by the increasing availability of astronomical data from telescopes like the Hubble Space Telescope and the James Webb Space Telescope. The sheer volume of data generated by these instruments is immense, demanding advanced computational techniques for analysis. Astronomers utilize sophisticated algorithms to identify and catalog celestial objects, analyze their light spectra to determine their chemical composition and distance, and simulate the evolution of galaxies and other cosmic structures. Counting the photons captured by these telescopes, analyzing their distribution across different wavelengths, and comparing these observations to theoretical models is crucial to expanding our understanding of the universe. This process underscores the crucial role of precise counting and sophisticated data analysis in astronomical research, revealing the vastness and complexities of the cosmos.

Climate modeling offers another compelling example of the convergence of advanced counting techniques and scientific understanding. Accurately predicting climate change requires the integration of vast amounts of data from various sources, including weather stations, satellites, and oceanographic buoys. These datasets, comprising temperature, precipitation, wind speed, and ocean currents, are processed using complex climate models that incorporate physical laws and numerical techniques. These models involve counting and averaging numerous variables over vast spatial and temporal scales to simulate the interactions between the atmosphere, oceans, land surfaces, and ice sheets. The reliability of climate predictions depends critically on the accuracy of these data and the sophistication of the models, emphasizing the significance of robust counting methodologies in addressing this critical global challenge. Furthermore, the accuracy of these models depends critically upon the integration of data from multiple sources, requiring sophisticated techniques for data aggregation, comparison, and validation. This process itself involves a substantial amount of counting and statistical analysis.

The field of finance relies heavily on advanced counting methods and sophisticated statistical techniques for risk management, algorithmic trading, and fraud detection. Financial markets generate massive amounts of data concerning stock prices, trading volumes, and economic indicators. Quantitative analysts (quants) use advanced statistical models and machine learning algorithms to analyze these data, identifying patterns and making predictions about market trends. These models often involve complex calculations and estimations, including counting and weighting diverse factors to assess risks and opportunities. For example, sophisticated algorithms are used to predict credit risk by evaluating vast datasets of financial transactions and credit histories. These analyses rely heavily on the ability to count and weigh various factors to assess the probability of default. Similarly, algorithmic trading employs sophisticated algorithms to execute trades at optimal prices, constantly scanning markets and adapting to changing conditions. This process involves rapid data analysis, including counting and comparing prices and volumes to generate trading signals.

However, this increased reliance on data-driven approaches also presents significant challenges. One key challenge is the potential for bias in data and algorithms. If the data used to train an algorithm is biased, the algorithm will inevitably perpetuate and even amplify those biases, potentially leading to unfair or discriminatory outcomes. For instance, algorithms used in criminal justice applications might exhibit bias if the training data disproportionately represents certain demographic groups. Addressing this challenge requires careful consideration of data collection and preprocessing techniques, including rigorous statistical analyses to identify and mitigate bias. This often involves detailed counting and comparisons of data subsets to ensure equitable representation. Furthermore, the ethical implications of using algorithms to make decisions that affect individuals' lives necessitate careful evaluation and transparent accountability, underscoring the critical importance of rigorous and unbiased data analysis in shaping fair and equitable outcomes.

Another challenge lies in the interpretability of complex models. While sophisticated algorithms can achieve high accuracy in making predictions, their internal workings may be opaque, making it difficult to understand why a particular decision was made. This "black box" nature of many algorithms poses challenges for accountability and trust. Developing methods for explaining the decisions of complex models is crucial for building confidence in these systems and ensuring responsible use. This includes the development of new methods to unpack the decision-making processes within complex algorithms, often involving techniques to analyze the weights and contributions of different variables in the model, thus highlighting the interplay between complex computations and human understanding.

Furthermore, the sheer volume of data generated in the 21st century presents considerable challenges for data storage, management, and analysis. The exponential growth of data requires the development of efficient and scalable data management systems, as well as sophisticated algorithms for efficient data analysis. Addressing these challenges necessitates not only advancements in computing hardware and software but also improvements in data visualization and communication techniques to make complex data accessible and understandable to a wider audience. This requires bridging the gap between the technical details of data analysis and the broader societal implications of using this data to drive decision-making.

In conclusion, the future of counting in the 21st century is inextricably linked to the advancement of technology and our ability to manage and interpret ever-increasing quantities of data. While new technologies and sophisticated algorithms offer unprecedented opportunities for understanding the world around us, they also present significant challenges related to bias, interpretability, and data management. Addressing these challenges requires a multidisciplinary approach, combining expertise in mathematics, computer science, social sciences, and ethics to ensure that the power of counting and data analysis is used responsibly and for the benefit of all. The journey of counting continues, evolving alongside our technological and societal advancements, requiring ongoing critical reflection to harness its transformative power while mitigating its inherent risks.

Counting and the Human Experience:
A Concluding Perspective

The preceding chapters have charted a remarkable journey, tracing the evolution of counting from its nascent stages in prehistory to its sophisticated applications in contemporary science and technology. We've witnessed the emergence of counting systems in ancient civilizations, their vital roles in agriculture, commerce, and the burgeoning fields of astronomy and navigation. We've explored how counting underpinned the scientific revolution, fueled technological advancements, and continues to be indispensable in diverse fields like genomics, astrophysics, climate modeling, and finance. But the story of counting is not solely a narrative of technological progress; it's deeply intertwined with the human experience, shaping our understanding of the world and our place within it in profound ways.

One of the most significant aspects of counting lies in its capacity to impose order on the seemingly chaotic universe. The act of assigning numbers to objects, events, or phenomena creates a framework for understanding patterns, relationships, and quantities. This ability to quantify and categorize is fundamental to scientific inquiry, allowing us to formulate hypotheses, test theories, and build predictive models. Without the ability to count, our scientific understanding of the world would be severely limited. Consider, for example, the field of cosmology. Our understanding of the vastness of the universe, the distribution of galaxies, and the expansion of space-time is fundamentally dependent on our ability to measure astronomical distances, count celestial objects, and analyze their properties through quantitative analysis. Similarly, in the microcosm, the understanding of the fundamental constituents of matter and their interactions relies heavily on precise measurements and the quantification of subatomic particles and their properties.

Beyond the realm of science, counting has played a pivotal role in shaping human culture and society. Early forms of counting facilitated the development of agriculture, enabling societies to manage resources, plan for the future, and establish complex social structures. The emergence of writing systems was closely linked to the need for record-keeping, a task that intrinsically involved counting and numerical representation. Throughout history, counting systems have reflected cultural values and beliefs, often serving as powerful symbols in religious and social practices. The development of calendars, for instance, is an intricate example of counting's interplay with cultural practices and belief systems. The cyclical nature of time, reflected in the structure of calendars, necessitated the development of sophisticated counting systems capable of tracking seasons, years, and even longer cycles, a task deeply intertwined with religious and agricultural practices.

The philosophical implications of counting are equally profound. The very act of counting presupposes the existence of discrete units, implying a certain degree of separability and individuality within the world. This concept of discreteness underpins much of our scientific thinking, leading to the development of models and theories that often rely on quantifiable variables and discrete entities. However, the relationship between counting and reality is not always straightforward. The act of counting can itself shape our perception of reality, leading to a certain degree of abstraction and simplification. For instance, the measurement of something like happiness or the assessment of environmental impact necessitates the conversion of complex realities into quantifiable indicators. These indicators inevitably involve a degree of approximation, highlighting the interplay between measurement and the subjective interpretations derived from such quantification.

The limitations of counting are also important to consider. Counting works best when dealing with discrete objects, but it struggles with continuous phenomena like time, space, or temperature. While we can measure these continuously changing parameters, the process inherently involves an act of discretization, converting continuous phenomena into manageable discrete units. Such conversion always introduces a degree of approximation, which in turn limits the precision and accuracy of our understanding of the phenomena involved. This inherent difficulty in applying counting to continuous data underscores a fundamental limit to the explanatory power of purely quantitative approaches.

Furthermore, the act of counting can be influenced by cultural, social, and political factors. The choice of units, scales, and methods of measurement can reflect underlying biases and assumptions, influencing the results obtained and their interpretation. Consider, for example, the historical measurement of national income or societal progress. Different metrics and methods of accounting can reflect different values and priorities, influencing our perceptions of economic growth or societal well-being. This underscores the importance of critical evaluation of data and methods in ensuring objectivity and validity in our understanding of the world.

The development of sophisticated statistical methods and computational tools has further expanded the possibilities and challenges of counting in the modern era. The ability to handle massive datasets, run complex simulations, and construct sophisticated statistical models has revolutionized many scientific and technological fields. However, this technological advancement also brings ethical considerations into sharp relief. The potential for bias in algorithms and data, the challenges of interpretability in complex models, and the need for responsible data management are critical issues that must be addressed. The application of counting and data analysis needs constant reassessment to maintain integrity, avoiding biases and mitigating the potential for unintended consequences.

Looking to the future, the interplay between counting, technology, and human experience is likely to become even more complex and multifaceted. Advancements in artificial intelligence, machine learning, and quantum computing will undoubtedly continue to expand the capabilities of counting and data analysis, opening up new frontiers of scientific discovery and technological innovation. However, the ethical implications of these advancements must be carefully considered, ensuring that these technologies are used responsibly and equitably. The ongoing development of robust methodologies for data validation and verification, alongside a critical examination of the potential biases embedded in algorithms and data, will be crucial to maintaining the integrity and reliability of this rapidly evolving field. Moreover, fostering a public understanding of quantitative reasoning and statistical literacy is essential to ensure that the power of counting and data analysis is used for the benefit of all.

In conclusion, the history of counting is not merely a chronicle of mathematical progress; it's an integral part of the human story. From the earliest tally marks to the sophisticated algorithms of the 21st century, counting has been a fundamental tool for understanding the world, shaping our societies, and driving technological innovation. As we move forward, the ethical considerations surrounding the use of data and algorithms must be paramount. A balanced and reflective approach is required to harness the transformative potential of counting while mitigating the inherent risks and ensuring the equitable application of its power. The journey of counting continues, interwoven with the ongoing evolution of humanity itself, demanding a sustained engagement with its transformative potential and its ethical challenges. The future of counting lies not only in further technological advancements but also in our ability to use this fundamental tool responsibly and ethically, fostering a more informed, equitable, and sustainable future for all. The story of counting is, in essence, the story of human ingenuity, and its continued evolution is witness to our persistent quest to understand and shape the world around us.

Appendix A: Sources and Further Reading

- Georges Ifrah, *The Universal History of Numbers: From Prehistory to the Invention of the Computer*

- Tobias Dantzig, *Number: The Language of Science*

- David M. Bressoud, *A Radical Approach to Lebesgue's Definition of Measure*

- Roger L. Cooke, *The History of Mathematics: A Brief Course*

- Online resource: *History of Mathematics* – MacTutor History of Mathematics Archive (https://mathshistory.st-andrews.ac.uk/)

- Documentary: *The Story of Mathematics* (BBC, available on YouTube)

Appendix B: Glossary

Term	Description
Abacus	An ancient calculating tool consisting of a frame with rods and movable beads, used for arithmetic operations.
Algorithm	A step-by-step procedure or formula for solving a problem or performing a task, especially by a computer.
Aqueducts	Engineering structures built to transport water from a source to a distribution point, crucial in ancient civilizations for supplying water.
Astrolabe	An ancient navigational instrument used to measure the altitude of stars and planets, aiding in determining latitude.
Binary System	A numerical system using only two digits, 0 and 1, fundamental to digital computers and data processing.
Cartography	The science and art of designing and producing maps, combining technical and aesthetic elements.
Chronometers	Precision timekeeping devices, crucial for determining longitude at sea and used in navigation.
Cipher	A method of transforming text to conceal its meaning, used in cryptography for secure communication.
Counting System	A systematic way of assigning numbers to quantity, such as decimal or binary systems.
Cuneiform	One of the earliest systems of writing, developed by the Sumerians, using wedge-shaped marks on clay tablets.

Decimal System	A base-10 number system using the digits **0** through **9**, the most widely used numerical system today.
Discrete	Refers to elements that are distinct and separate, often used in mathematics to describe countable sets.
Geodesy	The science of measuring and understanding the Earth's geometric shape, orientation in space, and gravity field.
GIS	Geographic Information System: A system designed to capture, store, manipulate, analyze, and manage spatial or geographic data.
Hieroglyphs	A system of writing using symbols or pictures, used in ancient Egypt for formal inscriptions.
Hindu-Arabic Numerals	The decimal numeral system originally developed by Indian mathematicians, later transmitted to Europe via the Arabs, now the most common numeric system used worldwide.
Metron	An ancient Greek term for measure, referring to tools or units used for measurement.
Number Base	The number of unique digits, including zero, used to represent numbers in a positional numeral system. (e.g., base 10 for the decimal system)
Sextant	A navigational instrument for measuring the angle between celestial objects and the horizon, essential for determining latitude and longitude.
Zero	A numeral representing no quantity or null value, serving as a placeholder in positional systems and enabling complex arithmetic operations.

About the Author

Jesse Luckett has had a highly fulfilling career as a computer electronics and software designer and engineer. Despite his retirement, he continues to maintain a profound passion for systems and applications design and development. Over his career, he has worked with a diverse array of hardware platforms, spanning from mainframes to personal devices. Jesse has developed systems and applications employing a wide range of software languages, including early assemblers, FORTRAN, and C, as well as contemporary languages such as C/C++, Java, Python, BASH, SQL/NoSQL, and others. His interests extend beyond the technical domain, encompassing the broader fields of science, technology, and engineering.

COUNTING